# Fort Monmouth

## The US Army's House of Magic

Melissa Ziobro

**BROOKLINE**
books
*Havertown, Pennsylvania*

Brookline Books is an imprint of Casemate Publishers

Published in the United States of America and Great Britain in 2024 by
BROOKLINE BOOKS
1950 Lawrence Road, Havertown, PA 19083, USA
and
47 Church Street, Barnsley, S70 2AS, UK

Copyright 2024 © Melissa Ziobro

Paperback Edition: ISBN 978-1-955041-22-5
Digital Edition: ISBN 978-1-955041-23-2

A CIP record for this book is available from the British Library

All rights reserved. No part of this book may be reproduced or transmitted in any form or by any means, electronic or mechanical including photocopying, recording or by any information storage and retrieval system, without permission from the publisher in writing.

Printed and bound in the United Kingdom by CPI Group (UK) Ltd, Croydon, CR0 4YY
Typeset in India by DiTech Publishing Services

For a complete list of Brookline Books titles, please contact:

CASEMATE PUBLISHERS (US)
Telephone (610) 853-9131
Fax (610) 853-9146
Email: casemate@casematepublishers.com
www.casematepublishers.com

CASEMATE PUBLISHERS (UK)
Telephone (0)1226 734350
Email: casemate@casemateuk.com
www.casemateuk.com

*Cover images:*
Front: Training at Camp Vail, 1920.
       Soldiers with equipment, 1995.
       "Potpie" gets his discharge from Colonel Clifford A. Poutre as the Army prepares to do away with the pigeon service.
       Fort Monmouth and its sub-installations like Camp Evans earned a reputation as the "Black Brain Center" of the United States, 1943.

Back: Troop review, Fort Monmouth, 1969.

All images courtesy US Army Communications-Electronics Command Historical Office, Aberdeen Proving Ground, Maryland

## Dedication

Dedicated to the men and women, military and civilian, who worked at Fort Monmouth, New Jersey 1917–2011—and the communities who supported them from start to finish.

# Contents

Foreword by Major General Randolph P. Strong     vii
Preface and Acknowledgements     ix
Introduction     xv

| | | |
|---|---|---|
| 1 | The Racetrack Paves the Way | 1 |
| 2 | "A Big Farm for Soldiers": The Great War Comes to Central New Jersey | 15 |
| 3 | The Inter-War Years: Radar and Other Research and Development Revelations | 35 |
| 4 | "Should They Fail, Expect Plenty of Hell": Training and Equipment Critical to Winning The Good War | 57 |
| 5 | "Where the Army Signal Corps Thinks Out Some of the Nation's Crucial Defenses": Cold War Battleground | 77 |
| 6 | "The Black Brain Center of the United States": Dr. Walter McAfee and His Colleagues Break Barriers at Fort Monmouth | 97 |
| 7 | "The Eyes, Ears, and Voice of the Fighting Man": The Vietnam War and Its Aftermath | 115 |
| 8 | "We Were Able to Control our Force Much Better than the Enemy Controlled His": Operations Desert Shield and Desert Storm | 131 |
| 9 | "It Has Definitely Left Some Lasting Marks that Aren't Necessarily Easy to Get Over": Support of the Global War on Terror | 147 |

10   The Army Goes Rolling Along: Behind the Scenes
     of a Base Closure                                    161

*Conclusion*                                              181
*Endnotes*                                                183
*Select Bibliography*                                     214

# Foreword

Fort Monmouth is a magical place, and nobody can tell its story better than Melissa Ziobro. Serving as the Command Historian for the United States Army Communications-Electronics Command (CECOM) she personally witnessed the Fort Monmouth workforce in action during major conflicts in Iraq and Afghanistan, while simultaneously conducting oral histories and maintaining a treasure trove of patents, documents, and artifacts in the command's archives. Simply, if it's part of Fort Monmouth's history, she knows all the details.

It's a story of dedicated, inventive, and brilliant people and the technologies they imagined, developed, and provided to not only the US military, but also to all of society. These technologies helped the United States win its wars for close to a century and made life easier and more enjoyable for all today. I would venture to say that every day we touch upon something that was imagined, developed, or enhanced at Fort Monmouth. Although primarily focused on supporting the US military, the extraordinary accomplishments of the Fort Monmouth workforce significantly impacted commercial industry and the everyday lives of all of us. When you listen to FM radio, think of Fort Monmouth's Major Edwin Howard Armstrong, who is widely considered the father of FM radio. When you fly across the country closely monitored by air traffic radars, think of Fort Monmouth's Colonel William Blair—often called the father of US radar. When you use satellite enabled communications of any kind, think of Fort Monmouth's Colonel John H. DeWitt Jr. and Project Diana, which proved you could send radio signals through the ionosphere and vacuum of space. Or you might think of Project SCORE (Signal Communications by Orbiting Relay Equipment), the program that developed the first satellite designed for relaying communications through space. The United States military is the most formidable warfighting force in history largely because of its technological superiority and much of that came from Fort Monmouth and the amazing people that worked there—and the local community that supported them.

Pro Patria Vigilans<br>
Randolph P. Strong<br>
Major General, US Army (Retired)<br>
The last Commanding General of Fort Monmouth

# Preface and Acknowledgements

This book will be published 20 years after I started working at Fort Monmouth in March 2004 as a civilian historian employed by the Department of the Army. This Army post in central New Jersey came into existence during World War I. Though it was initially supposed to be temporary, it wound up outliving the conflict. Until its closure in 2011, its various schools trained thousands of men and women for war, and its laboratories innovated countless technologies that saved lives on the battlefield and changed the world we all live in.

As the Army prepared to close Fort Monmouth in 2011, I chose not to move with my position to Aberdeen Proving Ground, Maryland for personal reasons. Even as I packed up the archive I worked in to ship it to Maryland, I vowed that I would keep Fort Monmouth's memory alive. Though I transitioned to working in academia, I've written newspaper and journal articles about the post; visited countless libraries and classrooms and historical societies to talk about its history; and have volunteered my time to install Fort Monmouth exhibits at the InfoAge Science and History Museums at InfoAge, a non-profit science and history museum located at a former Fort Monmouth sub-post and now National Historic Landmark, Camp Evans. Clearly, I agree with the World War I veteran of the post who said, "The place sort of gets into your blood."[1]

So when invited by Brookline Books to pitch a book proposal, I knew I wanted to do something Fort Monmouth related. But which part of Fort Monmouth history to focus on? Fort Monmouth's story is expansive, and complex. Most military posts have many "tenants," or military organizations, on them, and those tenants might be working on totally different missions. And the tenants come and go. You could write a whole book solely focused on United States Army Signal Corps training being located at Fort Monmouth from the World War I era through the Vietnam War era, or about the totally unrelated United States Army Chaplain Center and School being located there from 1979 to 1995, or the United States Military Academy Preparatory School (also known as West Point Prep) being located there from 1975 to 2011. You could write a whole book on the winged warriors of the United States Army Pigeon Service (always a crowd favorite), headquartered at Fort Monmouth from the World War I era until 1957. I could go on and on. The Annual Command Histories that my colleagues and I wrote each year, primarily focused on the post's support to operations in Iraq and Afghanistan, were each several hundred pages—several hundred pages for **one year** of history! But Brookline Books didn't ask for a series, they asked for a book! I decided that, in the interest of length, I'd tell a series of stories, loosely chronologically arranged, centered on Fort Monmouth as the "Army's House of Magic," as the *Saturday Evening Post* called it in 1952; a "big place that does big and bewildering things," like help pioneer air to ground radio, and radar, and communications satellites, and countless other technologies that at one point in time seemed too far-fetched to be real.

Having personally spoken to countless Fort Monmouth employees over the years, and read testimonials from folks who worked there dating back to 1917, I think the "House of Magic" title applies not just to the wizardry happening in the post's laboratories, but also the magical spell the post seems to have cast over people of every era. They love the buildings, the grounds, its close proximity to both New York City and Philadelphia, the surrounding communities, the nearby beaches, and so much more. You consistently find those who lived and worked at Fort Monmouth celebrating things like that fact that, "You can be at the beach in 15 minutes; you can be in the mountains in an hour"[2] and "I got to visit the sea coast, swim in the Atlantic Ocean, go and see New York City on leave."[3] They'll enthusiastically declare things like "there are few more attractive and comfortable camps anywhere …"[4] They'll call it "the general's paradise"[5] and the "playground of the Army."[6]

The focus of the book narrowed down, more or less, to Fort Monmouth innovations, the daring folks who dreamed them up and used them, and the deep and abiding love the post inspired in those who worked there, I next pondered how to ensure I was telling the story in a way that was really accessible and engaging. When, as a historian on post, I was tasked with anything from a research request about early night vision technologies, to writing an Annual Command History about twenty-first-century operations in Iraq and Afghanistan, I often felt as if the reports and other primary source materials I mined were written in a different language! A visitor to a 1960s open house at the post wrote on a comment card that the technology displays were "Interesting, but above the average person's head." Another comment card read, "Very interesting—a little too advanced for a layman."[7] Even Brigadier General George H. Akin, outgoing Deputy to the Commander of the United States Army Communications-Electronics Command (CECOM) at Fort Monmouth, noted in 1988, "the shocking thing that I had to get used to when I got here was the language ... and I think that is one of our problems. In CECOM and Fort Monmouth we speak our own electronics language and the rest of the world ... they don't understand us well enough."[8] In this book, I endeavor to explain the fascinating history in a concise way that you don't have to be an engineer to understand.

Then there's the fact that the Army loves organizational restructuring. When our story begins in earnest during World War I, we are talking about the United States Army Signal Corps' presence at Fort Monmouth. In the simplest possible terms, the Signal Corps is the branch of the Army created just prior to the American Civil War and charged with all manner of military communications.[9] More tenants moved onto the post as the years went on, as noted above. And the Army underwent several complicated reorganizations over the years, which meant the primary tenant at Fort Monmouth ceased to be the Signal Corps but rather a succession of different organizations with acronyms like ECOM (the Electronics Command), ERADCOM (the Electronics Research and Development Command), CORADCOM (the Communications Research and Development Command), CECOM (the Communications-Electronics Command), and so on and so forth. These organizations had increasingly complex organizational charts. As Colonel Raymond E. B. Ketchum II, CECOM's Deputy Commander for Resources and Management, said in 1987 of the organizational structure at Fort Monmouth, "Not in my wildest imagination did I expect to run into that kind of organizational mish mash."[10] I don't think the average reader wants to get bogged down in that "organizational mish mash," so I throughout the book I am going to simplify that piece as much as possible as the missions of these primary tenants remained more or less the same: ensuring the nation's Army, and often its military more broadly, had the electronic tools they needed to succeed on the battlefield.

Lastly, when writing this book I really wanted to use oral histories and other primary sources to let the past speak to the present in as unfiltered a manner as possible. As an educator, I am passionate about exposing my students to primary sources and I hope you enjoy being exposed to them here (minus having to dig through dusty documents). And as a former president of the non-profit professional development group Oral History in the Mid-Atlantic Region and someone who has done hundreds of oral history interviews over the course of their career, I believe firsthand accounts bring history to life in a way few other resources can. I want to center the words of the former workforce of Fort Monmouth as often as possible.

This book at times excerpts and builds upon my earlier work on the Fort, to include *Monmouth Message* newspaper articles no longer available online, journal articles, and a number of publications that were compiled and lovingly updated by generations of Fort Monmouth staff historians (myself included). These include *A History of Army Communications and Electronics at Fort Monmouth, New Jersey 1917–2007*; *A Concise History of Fort Monmouth, NJ*; and *Fort Monmouth Landmarks and Place Names*. While out of print, keen readers may be able to find copies of these publications online. The CECOM Historical Office archive now down at Aberdeen Proving Ground also contains a treasure trove of documents for serious researchers. Visits can be arranged by appointment.

I am deeply grateful to all of the Fort Monmouth historians past and present, to include Helen Phillips, Dick Bingham, Wendy Rejan, Chrissie Reilly, Floyd Hertweck, and Susan Thompson. Your tireless efforts to document and preserve the Fort's history mean folks can, for generations to come, continue to marvel at, and learn lessons from, the work done at the post. Fred Carl, founder of InfoAge, is a preservation hero who I hold in the highest regard ... Any royalties from this book will be donated to InfoAge in honor of my fellow historians who, through their past work, contributed to *Fort Monmouth: The US Army's House of Magic* posthumously or from afar.

Thank you to Mike Brady and Henry Kearney for being tremendous champions of the CECOM Historical Office, and to my academic mentors Dr. Katherine Parkin and Dr. Christopher DeRosa, for teaching me more than any curriculum chart ever required over the past 20+ years. I deeply appreciate all who provided feedback on portions of this manuscript, to include Fred Carl, Colonel (retired) James Costigan, Joseph Foster, Claire Garland, Floyd Hertweck, Rosanne Letson, Colonel (retired) Renita Menyhert, Lieutenant Colonel (retired) John Edward Occhipinti, Lee Ann Potter, Henry Kearney, Dr. Katherine Parkin, Wendy Rejan, and Chrissie Reilly (with apologies to anyone I may have forgotten). Finally, thanks to John, JP, and Donovan for enduring 20 years of never-ending "Did you know that Fort Monmouth ...?!" I love you!

# Introduction

Central New Jersey's Fort Monmouth worked to equip men and women in uniform with the latest electronics technologies from World War I through the twenty-first-century wars in Iraq and Afghanistan. The creations were often so innovative, so inconceivable to the general public, that the Fort had a mystical quality. For example, the *Saturday Evening Post* called Fort Monmouth the "Army's House of Magic" in August 1952, due to "the diversity of the activities" underway there.[1] The *Asbury Park Press* expanded on this theme in 1953, noting, "Over the years, military members and civilians from stenographers to scientists have contributed to development of the [Signal] Corps. The electronics magicians of Fort Monmouth's laboratories have many basic discoveries and inventions to their credit. Private research firms and industry are close partners, and the result of this combination has been the best available communications equipment for the man on the fighting front."[2]

How did this House of Magic develop? Chapter 1 predates World War I, providing context and explaining why the Army initially selected the site for Fort Monmouth. Monmouth County, New Jersey is a part of the ancestral homeland of the Lenape people. Its eastern edge touches the Atlantic Ocean and several rivers traverse the landscape, so the area's abundant hunting and fishing supported the Lenape—and then the European settlers who made their way to the "New World." By the time of the American Revolution, the colony of New Jersey was known as a powder keg or barrel tapped at both ends, nestled between bustling New York and Philadelphia.

Throughout the Revolutionary War, New Jersey was a hotly contested battleground. A recent report prepared in advance of the 250th Anniversary of the American Revolution reaffirms the oft-stated fact that the state can claim more than six hundred "battles, clashes, skirmishes and naval engagements, either fought in New Jersey or originating from New Jersey soil."[3] Of course, the nation gained its independence and the newly minted state grew and prospered.

While agriculture was the mainstay of the new state's economy, Monmouth County had sandy soil that could make farming tricky. Tourism would come to play a key role in supplementing the County's economy (as it still does). Scores of Gilded Age millionaires built sprawling "cottages" positioned where they could benefit from the sea breezes, and even those of the growing class of white collar workers in New York City made their way to the shore. Entertainments sprang up for the amusement of these guests, including the lavish Monmouth Park Racetrack and luxury hotel. The racetrack was billed as the largest in the world, and everyone who was anyone wanted to be seen there—even President Grant is said to have had a box.

The racetrack went out of business when a moralist movement convinced the state legislature to outlaw gambling. The site fell into ruin but would quite literally pave the way for the Army to come to central Jersey—the good roads in the area courtesy of the racetrack were a major selling point when the Army was scouting locations for a Signal Corps camp to train troops for World War I (along with access to water, a nearby train station, and relative proximity to the port of embarkation in Hoboken).

Chapter 2 explores how, when the United States entered World War I in 1917, it was fully recognized that the Army of some 200,000 was not going to make a decisive difference in Europe. An incredible mobilization occurred, and some four million Americans would ultimately serve in uniform. The Army's Signal Corps, with only roughly 55 officers and 1,500 men at the war's outset, was insufficient to furnish communications for the enormous Army that would be needed—and so the Signal Corps would obviously need to grow, too. As the Army began to search for land for additional Signal Training Camps, their investigation led them to Monmouth County. What would become the Fort Monmouth site was chosen after Carl F. Hartmann, then a lieutenant colonel and signal officer of the Eastern Department in New York City, tasked Charles H. Corlett, then a first lieutenant, to "go out and find an officer's training camp."[4] Corlett recalled his initial discovery of the site, formerly the home of the Monmouth Park Racetrack, in a 1955 letter addressed to Colonel Sidney S. Davis, Chairman of the Fort Monmouth Traditions Committee. He reported that after examining several other sites, he "finally stumbled onto the old Race Course near Eatontown. I found part of the old steel grandstand with eleven railroad sidings behind it, the old two mile straight away track and two oval race tracks, all badly overgrown with weeds and underbrush."[5]

Ultimately, the Army leased the land, some of which was being used as a potato farm, with an option to buy. On June 6, 1917, the *Red Bank Register* announced, "A Big Farm for Soldiers: Old Monmouth Park Leased by the National Government." The site would first be known as Camp Little Silver, for the town; and then Camp Alfred Vail, in honor of the New Jersey inventor who helped Samuel Morse develop commercial telegraphy. The camp prepared thousands for war, and its research and development laboratories worked on pioneering technologies like air to ground radio. The importance of battlefield communications like those facilitated by the camp cannot be overstated. As one doughboy noted, "When the hell begins, Signal contact becomes a man's lifeline. Without it, he is blinded."

The new post in New Jersey was a military success, contributing to the Allied victory. It was good for the state, too, as the sudden growth of the camp brought to the area a prosperity which had been absent since the height of the Monmouth Park Racetrack's popularity. The soldiers proved to be "good spenders" and "their relatives, sweethearts, and friends swelled the trade of the storekeepers."[6] Though the post was originally intended to close after the war, the Chief Signal Officer would instead authorize the purchase of the Camp Vail in 1919. Chapter 3 explores the growth of the post during the inter-war years. The Signal Corps School relocated to Camp Vail from Fort Leavenworth in 1919, and the Signal Corps Board followed in 1924. The installation would be granted permanent status and renamed Fort Monmouth in August 1925 (in honor of the soldiers of the American Revolution who fought and died at the nearby Battle of Monmouth). It would soon be known as the "Home of the Signal Corps."

The years between the World Wars saw tremendous growth at the post, including permanent construction that put local folks to work during the lean years of the Great Depression. While the military contracted following the Great War, and fewer men needed to be trained, the work of the research and development laboratories continued apace. Perhaps the most impactful development to emerge during the inter-war years was the first United States aircraft detection radar, the patent for which was awarded, postwar, to Fort Monmouth's Colonel William Blair. Fort Monmouth radar detected the incoming Japanese planes at Pearl Harbor, but the warnings of the radar operators were disregarded by their superiors. Fort Monmouth radar technician John Marchetti granted an interview some fifty years after the bombing in which he was asked, "When you heard about Pearl Harbor … how did you feel when you heard what had happened with the radar when the Japanese were attacking?" He replied, "I was furious. I was furious. The radar that had been built … was completely misused. It could have saved lives at Pearl Harbor; it could have changed the picture totally around. But instead it got us in the war, because that was the very real reason why we joined the war."[7]

Even before Pearl Harbor, President Roosevelt's proclamation of a state of limited emergency on September 8, 1939, following the outbreak of war in Europe, had mobilized Fort Monmouth. The Army was immediately authorized additional personnel, and from that point forward the number of men training at Fort Monmouth increased exponentially. Chapter 4 explores World War II on post. Tens of thousands of soldiers would ultimately train there during the War years at the Signal School, Officers Candidate School, and Replacement Training Center. Fort Monmouth laboratories worked frantically to make communications for these troops more transportable, more secure, and more reliable. Developments during this period included an early FM backpack radio that provided frontline troops with reliable, static-free communications, and near continuous advancements in the radar technologies first pioneered at the Fort in the 1930s.

The missions mushroomed such that the Signal Corps was required to purchase and/or lease numerous sites throughout the region to facilitate Fort Monmouth's expanded training and research and development missions, to include luxury beachfront hotels and farther flung sites like the Signal Supply Agency in Philadelphia. The expansion was necessary because the missions were critical. As one general in the field noted, "The Chief of Staff and myself have limited knowledge of Signal equipment. When Signal communications functions properly, as we expect they will, expect no praise. But should they fail, expect plenty of hell."[8] The troops trained and the technologies sourced out of Fort Monmouth would not fail, and would in fact contribute greatly to the Allied victory in what writer Studs Terkel famously called "The Good War."

The end of World War II did not necessarily mean the United States assumed a peacetime footing. The nation plunged right into the Cold War, and Chapter 5 explores Fort Monmouth on the front lines of that decades-long struggle for power with the Soviet Union (even employing former German scientist to help win a technical edge). The *New York Times* referred to the post as "Where the Army Signal Corps Thinks Out Some of the Nation's Crucial Defenses." This notoriety drew the attention of the infamous Senator Joseph McCarthy, who descended upon the post convinced a nest of Soviet spies operated on site. Dozens lost their jobs, as the post commander cooperated with McCarthy.

The space race is perhaps one of the more constructive elements of the Cold War, and Fort Monmouth personnel worked furiously on things like satellite technologies. When the Russians launched Sputnik I in October 1957, the Fort's Deal Test Area was the first government installation in the United States to detect and record the Russian signals. Dr. Harold Zahl, Fort Monmouth's Director of Research, later reported that "we had no legal project set up for tracking Russian satellites … but within our own laboratory, we had an immediate potential, and duty called desperately." He recalled that a "small select group of Signal Corps R&D personnel at Fort Monmouth," eventually dubbed the "Royal Order of Sputnik Chasers," vowed to work without overtime pay, twenty-four hours a day, seven days a week, "until we knew all there was to be learned from the mysterious electronic invader carrying the hammer and sickle."[9]

At times when the Cold War turned hot, as in the Korean War, Fort Monmouth's technologies again made a decisive difference. The introduction of automatic artillery and mortar locating radars in Korea proved to be a major success, helping soldiers to detect the source of incoming enemy attacks and to potentially launch counterattacks. Other developments of the period included field television cameras, pocket radiation detectors, and ever smaller, lighter, and more transportable field radios.

Advancements in communications and electronics systems came so far that in 1957 the Army discontinued the pigeon service, which had been a fixture on post since the end of World War I. "Hero" pigeons with distinguished service records were donated to zoos, others sold for $5 a pair. Newspapers across the country covered the end of the pigeon service, and folks lined up as far as the eye could see, hoping to acquire Fort Monmouth birds of their own.

The McCarthy episode was a dark period in the Fort's history with regards to civil liberties, but in the late 1940s and 50s, Fort Monmouth also earned a reputation as, in the words of African American electrical engineer Thomas E. Daniels, "the Black Brain Center of the US." Daniels and others affirmed that this post "provided a place where black scientists and engineers could find jobs and advance their careers," while other research facilities closed their doors to African Americans.[10] Chapter 6 explores this phenomenon. Of course, these employment opportunities did not inoculate the Fort's African American employees against the culture of discrimination and segregation that marked this period in our country's history. Keep in mind that the Army itself institutionalized discrimination until President Harry S. Truman signed Executive Order 9981 on July 26, 1948, ending segregation in the United States Armed Forces. Jim Crow ruled in the private sector, and New Jersey's Ku Klux Klan chapters marched openly in the streets into the 1920s. Despite its best efforts, the Signal Corps could not ensure uniformly equitable treatment for African American Army personnel. Examining the experiences of just a few of the Fort's early African American employees illustrates the dichotomy between the Signal Corps' relatively progressive hiring policies and the day-to-day lives of the African Americans benefiting from them in central New Jersey.

A key figure discussed here is Dr. Walter McAfee. Dr. McAfee surmounted the racism endemic to twentieth-century America and made unique and enduring contributions to the scientific community during his 42 years working as a government scientist at Fort Monmouth, to include critical contributions to Project Diana, which allowed the first "contact" with the moon in 1946. He also made time to quietly battle injustice, teach, and mentor a new generation of innovators and leaders as a professor at Monmouth College (now University) in West Long Branch, a trustee at Brookdale Community College in Lincroft, and an organizer of enrichment programs for high school students.

Even as the Korean War ground to a bloody stalemate, it was becoming clear to those following world affairs that Southeast Asia was a powder keg. As the French abandoned the area following their stunning loss at Dien Bien Phu, the United States become increasingly entrenched in supporting the South Vietnamese (in support of the domino theory). The number of American advisors on the ground ticked steadily upward, then gave way to massive bombing campaigns and ground troops in what would become a painfully divisive war in Vietnam. BG Walter E. Lotz, Assistant Chief of Staff for Communications-Electronics (ACSC-E), US Army, Vietnam (USARV), stated in 1966 that "Electronics has never been so vital in a war as it is here in Vietnam."[11] Chapter 7 explores how Fort Monmouth managed signal research, development, and logistics support, and supplied combat troops with a number of high-technology commodities during the Vietnam conflict. These included mortar locators, aerial reconnaissance equipment, sensors, air traffic control systems, night vision devices, and surveillance systems. Of course, despite the technological edge, Americans, like the French, ultimately withdrew from the quagmire in Vietnam, and Saigon fell in the end.

Although the end of the Vietnam War meant reduced training missions at Fort Monmouth, the post continued its research and development work throughout the 1970s, 80s, and into the 90s—as covered in Chapter 8. Military and civilian personnel based out of Fort Monmouth were therefore ready to work round the clock during the Gulf War to equip soldiers with everything from jammers to night vision, to surveillance and intelligence systems, and to sustain these systems in the field. Fort Monmouth's systems gave American forces unprecedented capabilities for communication, command and control, surveillance, target acquisition, fire control, position, and data analysis.

As Dr. Richard Bingham wrote in his 1994 monograph, *CECOM and the War for Kuwait, August 1990–March 1991*, "The roles these systems played in Operation Desert Shield and Operation Desert Storm attest to the significance of the technologies CECOM and predecessor organizations introduced to the battlefield in recent decades and the importance of CECOM support for these technologies during preparations for and pursuit of the Gulf War." The United States and its Allies won the war for Kuwait "not because they amassed a superior force with superior fire power, but because we were able to control our force much better than the enemy controlled his; because we were able to know what the enemy was doing while denying him knowledge of what we were doing; because we were able to see and target enemy forces before they could see us. These were the abilities that [Fort Monmouth]-managed commodities gave the Army."

Chapter 9 considers how, when the September 11 tragedy struck, Fort Monmouth personnel sprang into action, supporting rescue and recovery efforts at Ground Zero and the Pentagon. Fort fire personnel assisted with survivor decontamination. Bomb disposal specialists from the 754th Ordnance Detachment deployed to Ground Zero to see how they could assist. Engineers and contractor teams deployed to New York to help find survivors in the rubble by locating their cell phones and using tiny infrared cameras to search through voids in the rubble for signs of life. A laser doppler vibrometer developed at the Fort assessed the structural integrity of nearby buildings, helping to protect rescue workers from building collapses. One intelligence analyst from Fort Monmouth who deployed to Ground Zero recalled in an oral history interview, "oh yea, I've got PTSD, any time I hear fire trucks and ambulances ... it was very tough going in there, and it has definitely left some lasting marks that aren't necessarily easy to get over."[12] Fort Monmouth also deployed a quick reaction task force to the Pentagon to install a communications infrastructure for thousands of displaced workers there. The post then quickly pivoted to support the "Global War on Terror" that followed.

Fort Monmouth supported operations in both Iraq and Afghanistan, but the importance of its missions could not protect the post from closure. Chapter 10 gives a behind the scenes look at the ultimately losing battle to save Fort Monmouth from the Base Realignment and Closure Commission of 2005 recommendation, and how and why the Army decided to close Fort Monmouth and shift the bulk of the operations at Fort Monmouth to Aberdeen Proving Ground (APG), Maryland. It was not a sign that the missions conducted at Fort Monmouth were unimportant—just that the Army thought they could more efficiently be conducted elsewhere. Was that actually the case? Did "brain drain" threaten combat effectiveness? How did the post closure impact mission readiness, employee well-being, and the economies of New Jersey and Maryland? And what does the future hold for the site? Whatever that looks like, one hopes it includes a nod to the generations of life-saving and innovating "magicians" of Fort Monmouth's past.

CHAPTER I

# The Racetrack Paves the Way

The land that would eventually become Fort Monmouth is located in Monmouth County, New Jersey. It is the ancestral homeland of the Lenape people. Situated not far from the Atlantic Ocean and closer still to the Shrewsbury River, the area's abundant hunting and fishing were of great appeal to the Lenape—and then to European settlers as they made their way to the "New World." By the time of the American Revolution, the colony of New Jersey was known as a powder keg or barrel tapped at both ends, nestled between bustling New York and Philadelphia.[1]

Throughout the Revolutionary War, New Jersey was a hotly contested battleground. A recent report prepared in advance of the 250th Anniversary of the American Revolution reaffirms the oft-stated fact that the state can claim more than six hundred "battles, clashes, skirmishes and naval engagements, either fought in New Jersey or originating from New Jersey soil."[2] Of course, the nation gained its independence, and the newly minted state grew and prospered.[3]

While agriculture was the mainstay of the new state's economy, Monmouth County had sandy soil that made farming tricky. In the latter decades of the nineteenth century, tourism would come to play a key role in supplementing the County's economy (as it still does today). Scores of Gilded Age millionaires built "cottages" positioned where they could benefit from the sea breezes, and even those of the growing class of white collar workers in New York City made their way to the shore. Entertainments sprang up for the amusement of these guests, including a world-renowned horse racing track and luxury hotel. The racetrack complex was billed as the largest in the world, and everyone who was anyone wanted to be seen there—even President Grant was said to have purchased a box.[4]

The racetrack and hotel went out of business when a moralist movement led the state legislature to outlaw gambling. The site fell into ruin but would quite literally pave the way for the Army to come to central Jersey—the good roads in the area courtesy of the racetrack were a selling point when the Army was scouting locations for a Signal Corps camp to train troops for World War I.[5] Just how did Monmouth County become a summer tourism destination, and how did that lead to the Army's 90+ year residency in the area?

## Summering at "the Branch"

As noted above, much of the Monmouth County shore region is plagued with sandy soil that can make agriculture difficult. Thus, with the exception of "nucleated villages ... along major routes,"[6] the area remained sparsely settled until the rise of seashore resorts and tourism in the late nineteenth century, when the Civil War era and the wave of industrialization that followed brought prosperity to much of the North. Businessmen profited handsomely, and accumulated excess money. Many of these people were "new money" industrialists. They rented summer cottages in Monmouth County where they could take advantage of the sea breezes. Others outright purchased tracts of land and constructed magnificent summer homes.[7] Long Branch was especially popular.[8]

Concurrently, a new type of worker evolved. These salaried, white collar employees regularly received a week or more paid vacation per year. These scheduled, compensated vacations gave people the time and extra money to plan family trips.[9] This, combined with improvements in steamship and railroad transportation during the second half of the nineteenth century, allowed the Jersey shore to become a popular summer vacation retreat even for less affluent New Yorkers.[10] According to one historian, "The contrast between taking a ship and then a train to the shore (from NYC) as compared to bumping along in a coach was like day and night."[11] The *New York Times* declared, "We know of no more varied and satisfactory a jaunt to be had for less money."[12] These middle class folks might not be able to build second homes at the shore, but they could day-trip, thanks to quicker, modern transportation, or they could stay in one of the many hotels that sprung up to accommodate the demand. At popular hotels such as the West End, the Continental, the Clarendon, the United States, the Mansion House, the Pavilion, or the Metropolitan, a week's stay could cost anywhere from $20–$35 per person per week (when in 1890 the average annual income was $380 per year).[13]

The city of Long Branch gained such notoriety that it became the favored seaside resort of seven US Presidents: Ulysses Grant (1869–1877), Rutherford B. Hayes (1877–1881), James A. Garfield (1881), Chester A. Arthur (1881–1885), Benjamin Harrison (1889–1893), William McKinley (1897–1901), and Woodrow Wilson (1913–1921). It was even dubbed the "summer capital," due to the frequency of Grant's visits to his five-acre cottage in 1873.[14] A visit by the president could mean thousands of dollars in increased revenues per day in Long Branch.[15]

To combat a slight wane in tourism caused by competition from other resort towns like Saratoga, New York, and Newport, Rhode Island, some of the area's wealthier summer inhabitants clamored for the introduction of horse racing to the area. They strove to make the Jersey Shore the "Newmarket of America."[16] According to the *New York Times*, the "myriads of fashionable visitors unanimously expressed their conviction that to make it the most attractive and delightful of water places, a race-course and racing, properly and decorously managed, was the only desideratum."[17] Among the entrepreneurs and politicians leading the charge were businessman—and avid gambler—John F. Chamberlin (sometimes spelled Chamberlain) and New Jersey Senate President Amos Robbins.[18]

## The Original Monmouth Park

Chamberlin was inspired to construct the Monmouth Park racetrack while on a foxhunt in the area in 1865.[19] He and his partners purchased 128 acres of the Corlies Estate in 1869. The land, located three miles from Long Branch in Eatontown, included a residence, a barn, and a wagon house. It was a two-and-a-half-hour trip from New York or a three-hour trip from Philadelphia. Davison and Chamberlain fenced the grounds and laid out an oval, 80-foot-wide, one-mile racetrack that opened on July 30, 1870.[20]

This park was in what would later be the southern portion of Fort Monmouth, in the vicinity its Patterson Army Health Clinic (for those familiar with the area).[21] The arched, wooden entrance was located on today's Broad Street, near Park Avenue.[22] In season two steamboats, or "floating palaces," made daily runs from Pier 28 in New York to Sandy Hook. There, patrons could make a connection to the park by rail.[23] The round trip price was under two dollars.[24] It was said that from the 400 foot long, 7,000 seat grandstand every foot of the course could be seen. This was a trait that other tracks, such as Jerome Park in New York, lacked.[25]

Opening day was a slight disappointment despite the fact that the *Jesse Hoyt* and *Plymouth Rock* steamships brought hundreds of people from New York City to Sandy Hook to get the train to Long Branch. Financiers Jim Fisk and Jay Gould, as well as the infamous Boss Tweed, were among those present. While "Tammany politicians were as plentiful as they are on the eve of an election around the precincts of City Hall," at least 2,000 of the grandstand seats went empty.[26] President Ulysses S. Grant was noticeably absent, though he was known to vacation in the area.

That first season lasted just five days, with races on July 30 and August 2, 3, 4, and 5. Races were sponsored by local hotels and businessmen, with a total of $31,000 in purses posted. Boss Tweed even donated money for the "Tweed Purse."[27] The *New York Times* declared the overall opening "highly credible ... though not crowded."[28]

While the 1871 season was declared "fair, though not flattering,"[29] it is said that President Grant eventually capitulated and purchased a box. The masses soon followed suit. By 1872, a record crowd of 25,000 watched the Monmouth Cup Race.[30] Racegoers filled hotels to capacity.[31] By 1873, it was said by the *New York Times* that Monmouth Park denied the famous Saratoga Resort "her fair chance" at making a profit during racing season.[32]

The races even offered a measure of sexual equality in the stifling Victorian era. Betting by females was acceptable at Monmouth Park, whereas in hotels and betting parlors it was frowned upon.[33] The methods women used to make bets were usually dismissed as whimsical. Women were often accused patronizingly of "betting upon their sympathies in a very charming and decidedly feminine fashion."[34] While the *New York Times* did concede that "some of these (women) who can overcome the peculiar prejudices and whims which so often control a woman's judgment, have been very successful as betters, and can count their winnings from a small original investment by thousand," they generally maintained that:

> Their entrance into the fascinating game is attended by many queer notions. They are ignorant as to the merits of horses and riders. The novitiate then looks over a list of horses, and finds a name that pleases her, and places her first $5 in the mutual pool box to bank a horse whose name she fancies. Old betting men say that Caramel has probably been more favored in this way than any other horse on the turf, and hundreds of women have lost their first $5 dollars because that is "such a sweet, pretty name." The women learn better after a while. Another stumbling block for the novices are the colors sported by the jockeys. A favorite color on a spruce jockey often draws out from the pocketbook of some fair one the first $5 she has ever bet on the uncertain chances of a dash around the track. If she wins the color ever remains her favorite. If she loses that color also loses for her all attractiveness it may have before had. Once a woman is "in the swim" of the running track, however, she will remain there just as long as her money holds out, and even then the fascination of the track clings to her and she is ever its slave.[35]

These women, often of high esteem in the community, could hire escorts to accompany them to the tracks. The escorts would help them find their seats, offer betting tips, provide physical protection, and purchase the women's tickets for them. Most importantly, according to sources, attending the races with an escort as opposed to a husband secreted frivolous gambling expenditures. According to one escort:

> Women, you know, have the betting fever just as badly as we do, and some of those who have it the worst couldn't gratify it if it wasn't for fellows in my profession. You see, some of them play the races on the sly. They don't have their husbands to bring them to the tracks, and even if they did they wouldn't bet as much as they do, for their husbands wouldn't let them. So they hire escorts to go to the races with them.[36]

The escorts were required to have "good address, a fair education, a thoroughly controllable temper ... a good knowledge of turf events, the breeding of horses and their performances, and familiarity with the merits and demerits of jockeys. He must always have good clothes and seem the gentleman escort rather than the professional." They earned around two dollars a day plus expenses.[37]

The park changed ownership in 1878 when a company headed by George L. Lorillard purchased it. That year, a race could easily sell as many as 13,500 tickets for a total gross of at least $69,880. In 1882, the longest racing season to date took place. Gamblers avoided Saratoga, Sheepshead Bay, and Jerome Park in favor of Monmouth. Not even Coney Island threatened the track's popularity.

The New Jersey Central and Pennsylvania Railroads reorganized themselves to accommodate racetrack-goers, running trains over the same track between their terminal at North River and Long Branch. A round trip steamboat ride from New York to Long Branch cost sixty cents; the round trip ferry ride from Coney Island, thirty cents; and the round trip railroad ride, one dollar and fifty cents (with that price expected to drop in order to remain competitive). The year 1885 saw crowds larger than ever reported at any US racetrack. By 1888, purses had jumped from $12,600 to $210,850. The track became so popular that expansion was a necessity.[38]

## Expansion

A bigger, fancier Monmouth Park was built just north of the original track. Designed by David D. Withers, it opened on July 4, 1890 and featured a one-and-one-half mile oval track, centered on what later became Fort Monmouth's Greely Field; a one-mile straight-of-way; a 700 by 210 foot steel grandstand for 10,000 spectators (reputedly, the largest in the world); and a luxury hotel, fronting Parkers Creek. The new park was three times the size of the original, and encompassed 640 acres—almost all of what would become Fort Monmouth's "Main Post." It was touted as the largest racetrack complex in the world.[39]

The railroads again made adjustments to accommodate the racetrack. According to newspaper reports:

The railroad facilities will be even better than they were at the old track, and those were the best in America. The railroad sidings reaching the track are a show in themselves. They are eighteen in number, and the most remote one is nearer to the main entrance to the grandstand by one half than was the nearest one at the old track. The turntable used in connection with them is the largest ever constructed in America. These facilities will enable the railroad companies to handle 20,000 people and get them away from the racetrack in a quarter of an hour after the conclusion of the races.[40]

"The exuberantly colorful" people of the era, such as financier and philanthropist Diamond Jim Brady, actress Lillie Langtry, and opera singer Lillian Russell, frequented the races and the Monmouth Park Hotel.[41] Other notable attendees included oil tycoon John Warne "Bet a Million" Gates; tobacco millionaire Pierre Lorillard IV; English poet Alfred, Lord Tennyson; boxer James Corbett; and Jessie Lewisohn of the banking family. Mike and Charlie Dwyer, whose stable included the world's finest thoroughbreds, hosted lavish suppers. The Drexel family of bankers from Philadelphia gave exclusive soirees. Even politicians like Governor Bowie of Maryland and Senator Stockton of New Jersey tried their luck at the races.[42] Newspapers across the country reprinted the exploits of these proto reality stars. One wonders what readers of the *Northern Pacific Farmer* thought when reading:

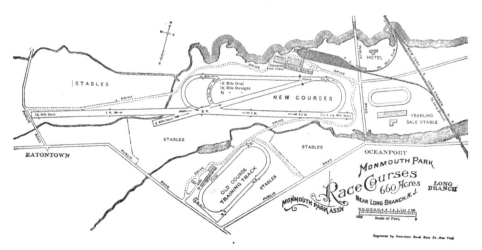

The original Monmouth Park opened in 1870; the new course, shown here, opened in 1890. (Courtesy US Army Communications-Electronics Command Historical Office, Aberdeen Proving Ground, Maryland)

> People have learned to think of [socialite] Freddie Gebhardt as a brainless dude … but he is in some of the ordinary transactions of life an astute and remarkably shrewd person. This is particularly made apparent in his connection with the turf. He is an enthusiastic attendant upon racing-courses, and owns in Eole one of the best long distance race-horses in the country, if not in the world. Usually young men with large fortunes who go on the turf retire at the end of a few seasons in a badly shattered condition financially. Gebhardt has proved a marked exception to the rule. Instead of being skinned he has assiduously devoted himself to the labor of removing cuticle from other people. Gebhardt's latest exploit occurred at the Monmouth Park race-track recently …[43]

That "latest exploit" was winning about $57,000 on one race—about $1.7 million dollars today.[44]

One of the most amusing anecdotes from the era is that of Lillian Russell riding to the track on a bicycle studded with diamonds. The bicycle was a gift from Diamond Jim Brady. Is it a true story? There are no photos to be found, so we may never know. It certainly seems plausible given the many documented excesses of the Gilded Age.[45] The duo also reportedly drove about in a custom-built electric automobile.[46]

The Monmouth Park hotel, on Parkers Creek, 1890. (Courtesy US Army Communications-Electronics Command Historical Office, Aberdeen Proving Ground, Maryland)

Reporters and artists (latter day paparazzi) from the popular periodicals of the day followed the moneyed and famed in order to capture exploits such as these.[47]

However, not all accounts of the racetrack were so glowing. One *New York Times* critique asked: "Is it really necessary ... that the food served at the so-called refreshment stands on the track and in the boats should be so very vile, that the plates should be so very dirty, and the waiters so slovenly and ill-trained?" This article ultimately concluded, however, that "even greater drawbacks could diminish but little the pleasure of this really delightful trip."[48]

There will always be critics. Still, the Monmouth Park experience proved so delightful that many people were reluctant to leave. Investors built the Monmouth Park Hotel on Parker's Creek to accommodate these attendees of the races. The massive building had 153 rooms. Amenities included an electric elevator, a smoking room, and a billiard room.[49] It was described as "the zenith of Victorian opulence with a surfeit of gold braid, silk tapestries, glass chandeliers, oriental rugs, and baroque staircases."[50]

## The End of an Era

The new racetrack was open for only one year when an anti-gambling faction began pushing more strongly for the end of legalized betting. The park had briefly closed in 1877 as the result of a law that "classed betting booths with disorderly houses," and by 1891 the Monmouth Racing Association had moved many of the races to Jerome Park.[51] According to the *New York Times*,

> This transfer will be regretted by very many people because of the hot and dusty trips to [other] tracks that will be necessary for people who remain in the city. The rapid and perfect service of the New Jersey Central and Pennsylvania Roads will be greatly missed, as will also the cool and pleasant nights at the Branch which have been so pleasurable to racegoers during the Monmouth Park season in past years.[52]

The transfer was also regretted by local farmers, merchants, and hotel owners because the "closing of the track near Long Branch last summer entailed a loss of hundreds of thousands of dollars to the various seashore interests."[53] The railroads alone were expected to lose $5,000 per day in the absence of racing at Monmouth Park.[54]

The racetrack reopened for its 1892 and 1893 seasons. Louisiana gambling magnate John A. Morris secured control of the track in 1893 after the death of Withers. Some 30,000 people (apparently were not concerned with the morality of racing) were estimated to attend the 1892 opening. The *New York Times* called Monmouth Park "a perfect race track, and far and away the best in the country."[55]

Shortly thereafter, though, Morris entered into a very public feud with the *New York Times* after that newspaper ran a series of scathing exposés about the supposedly fixed races and unbeatable odds at Monmouth Park. Morris banned reporters from the Park. At least one reporter in turn sued the Park for access. The *Times* doubled down, labeling Morris "the lottery king, who was spewed out of the state of Louisiana after he had prostituted that state to his purposes until the people would stand it no longer."[56] By July 1893, the same newspaper which had declared Monmouth a "perfect race track" one year earlier was warning readers that "the trail and slime of the lottery serpent about the race track is a little too much for decent people."[57] The *Times* was not alone in deriding the park's new management and clientele. Other papers waded into the fray. The *New York Tribune*, for example, declared in July 1893 "If the attendance at the Monmouth Park racetrack continues to fall off as it has … it will soon consist chiefly of track-owners, bookmakers, trainers, and what, for want of a more descriptive name, may be called ladies-in waiting … there was not a well-known face in the crowd …"[58] A Monmouth Park contingent sued select newspapermen for $100,000 for libel, to no avail.[59]

A "most salacious and cold-blooded" murder at the track that summer did nothing to help Monmouth Park's reputation. Well-known horse jockey and trainer Patrick S. Donovan, better known as "Snip," killed his longtime friend and colleague, John Chew, early on the morning of August 6. A drunken fight over money led to Snip (who was employed by the wealthy Pierre Lorillard) stabbing Chew. Snip originally pled not guilty to the charge of murder. When he retracted that plea and instead pled guilty to manslaughter, a local paper reported "the people who had gathered from

Racing program. (Courtesy George Moss Collection at the Monmouth University Murry and Leonie Guggenheim Memorial Library)

all over the County and many from out of the State were amazed and very much disappointed as they thought this case would be one of the most interesting trials on record." Snip, who suggested he receive a $1,000 fine, was instead sentenced to ten years in the state prison. (Snip would be released in 1895. One editorial called his early release "a case of mistaken clemency," noting, "Many men have hung on less incriminating evidence ... The Court of Pardons are either easily gulled or someone in whom they have confidence has imposed on them." He died of heart disease in 1902. At least one obituary failed to mention the murder, and hailed Snip as "one of the best known turfmen in the United States.")[60]

The allegations of corruption and bad behavior and the lawsuits with the newspapermen and the murder were a public relations nightmare as a "moralist movement" championed by state senator James A. Bradley gained steam and targeted the track.[61] The movement enjoyed bipartisan support, with one headline declaring, "Uniting Against the Gamblers: Republicans and Democrats Combine in Opposition to Monmouth Park."[62] The newspapers helped fan the flames, with one editorial noting that "tens of thousands of people" were talking about "the evils of racetrack gambling," and "the fact that Monmouth Park and the racetracks like it are so pervaded and influenced by the gambling element that they seem to exist rather as associations for the promotion of gambling than as organizations for the improvement of the breed of horses." That particular editorial concluded, "The betting

Monmouth Park. Circa 1893. (Courtesy US Army Communications-Electronics Command Historical Office, Aberdeen Proving Ground, Maryland)

ring appears to be paramount al Monmouth Park. Everything else is seemingly secondary."[63]

The movement led the New Jersey legislature to outlaw gambling once and for all in 1894. While many legislators tried to protect the tracks, they were overwhelmingly opposed and ultimately defeated by "ministers, priests, lawyers, and others." They were labeled "Race-Track Politicians"[64] and accused of making legislation

> a matter of purchase and sale ... In their greed the racetrack managers have overstepped all bounds of prudence and decency. They have antagonized the law abiding and peace loving citizens of the state. The festering sores these gambling aliens have created have been torn open and their putridity and other rottenness exposed.[65]

Compromises such as funneling five cents from every admission ticket sold to the state or confining gambling to certain areas such as the racetracks, and only on designated days, were discarded.[66] Anti-gambling proponents could not be swayed in their belief that gambling meant "the ruin of young men by the thousands ... the influx to this section of a base, pestilential multitude, the shame of all moral people; in short, the offering of ourselves as a swill barrel, into which New York, Philadelphia, and adjacent cities may empty their slops."[67] The Reverend S. Edward Young asserted that the "main magic that entices people to the track is the opportunity to gamble and get drunk, and more evil will be wrought at the track in eight weeks than all the pulpits can correct in fifty-two."[68]

The investors backing the track and its various races invested their money elsewhere—to include Kentucky (whose famed Derby was first run in 1875).[69] By September 1894, the *New York Daily Tribune* observed, "The very benches have been removed from the dismantled and ruined grandstand at Monmouth Park and all the racetrack property in New Jersey is utterly worthless."[70] In 1895, the Monmouth Park property was divided into four parcels and auctioned under foreclosure proceedings by the Farmer's Loan and Trust Company. While the Withers' estate held $384,000 of the park's mortgage, the syndicate that held the other one-sixth forced the sale. It took place at the courthouse in Freehold. There was little bidding. Lots sold for much less than their appraised value. Five hundred and ninety acres, including the original and new tracks, grandstand, clubhouse, and stables, went undisputed to Judge A. C. Monson (executor of the Withers' estate) and A. J. Cassatt for $50,000. Twenty acres, including the hotel, also went undisputed. Forty acres, including the yearling stables, went to Augustus Carson (Withers' nephew) after some less than spirited bidding against one Lucien Appleby.[71] The fourth and final plot of 4.5 acres, with two dwellings, went to Mr. Cassatt and Judge Monson for $2,500. At the auction's end, the entire property was held by the Withers' estate.[72]

Unfortunately for the estate, an 1897 constitutional amendment against gambling and bookmaking squashed any hopes of a Monmouth Park revival.[73] While horse racing itself was not outlawed, the "prohibition of gambling brought about the same result."[74] The "flush prosperity" of Long Branch was doomed.[75] The property was put up for auction at the Real Estate Exchange Rooms in New York City on April 22, 1897 by Adrian Muller and Sons. The decision to auction the property was made by the heirs of David Withers, who, according to the *New York Times*,

The ruins of the Monmouth Park grandstand, once touted as the largest in the world, 1899. (Courtesy US Army Communications-Electronics Command Historical Office, Aberdeen Proving Ground, Maryland)

"have probably concluded that there is no chance for racing over that magnificent piece of property in the near future."[76] The April 3, 1897, edition of *The Thoroughbred Record* reported:

> The news will be received with regret that it has been definitely decided to put up the magnificent Monmouth Park race track at auction on April 22, 1897. Up to the last minute, the owners and mortgagees had hoped for a turn in the tide of public sentiment, but it is doubtful whether during the next ten years any favorable amendments to the existing laws in the State can be urged with any fair chance of success.

Deserted, the grandstand, track, and hotel fell into ruin. A storm destroyed the grandstand in 1899, and the hotel burned to the ground in 1915 in a "fire of mysterious origin."[77] Much of the land would remain abandoned and underutilized until in 1917 the Army realized—not unlike the Monmouth Park racetrack investors—what a prime piece of real estate it was.

CHAPTER 2

# "A Big Farm for Soldiers"

## The Great War Comes to Central New Jersey

In the spring of 1914, Europe was a tinderbox. Countries vied for power, and nurtured festering resentments against each other. The June 1914 assassination of Archduke Franz Ferdinand, heir to the throne of Austria-Hungary, triggered a complex series of alliances that led country after anxious country in Europe to declare war on each other. As the military historian B. H. Liddell Hart wrote, "Fifty years were spent in the process of making Europe explosive. Five days were enough to detonate it."[1] On one side, Britain, France, and Russia formed the nucleus of what was known as the Triple Entente (also known as the Allies). On the other side, Austria-Hungary and Germany were the main players of the Triple Alliance (also known as the Central powers). Dozens of other countries were involved to varying degrees. The United States, Japan, and many other nations would join the Allied powers; the Ottoman Empire and Bulgaria would join the Central powers. Russia would withdraw to face a revolution at home.[2]

When World War I broke out in Europe in 1914, the combatants expected it to be a quick fight. The United States fully intended to remain neutral, and President Woodrow Wilson's administration worked quietly to try to help broker peace. Provocations such as the sinking of American ships by German U-boats attempting to blockade their enemies, though, pushed some Americans in the direction of war. This pro-war sentiment calmed a bit when Germany promised to stop submarine warfare against Americans. The eventual resumption of unrestricted submarine warfare, which sank many civilian American ships, and the Zimmerman Telegram from Germany to Mexico, suggesting an alliance between the two nations in the event of America's entry into the war, proved to be the last straw. On April 6, 1917, the United States declared war on the German Empire. They thus joined France, Great Britain, Russia, Canada, Australia, New Zealand, South Africa and Italy—"the Allies"—in their fight against Germany, Austria-Hungary, the Ottoman Empire, and Bulgaria. The United States faced a coalition war in which it would be, according to noted military historian Russell Weigley, a "military novice entering belatedly upon a deteriorating situation."[3]

The US Army of some 200,000 was far too small to make a decisive difference in Europe. As one historian noted, "Masses of serried troops fighting in trench warfare to be properly integrated with the forces of our allies would demand rapid and drastic changes in tactics, organization, and equipment."[4] An incredible mobilization would be needed. Ultimately, some 4 million Americans would serve in uniform. Roughly 2 million made it to France; some 1 million saw combat. This mobilization led directly to the birth of what would become Fort Monmouth.[5]

The Army's Signal Corps, with its strength of 55 officers and roughly 1500 men at the war's outset, was entirely insufficient to furnish communications for the enormous Army that would be raised. This branch of the Army had been organized by Albert J. Meyer in 1860 just prior to the American Civil War. Its initial duties centered largely on signaling with red and white "wig wag" flags, and manning aerial observation air balloons and telegraph machines. By the time the United States entered World War I in 1917, the Corps was integrating far more advanced technology like telephones, radio, and even airplanes.[6] The Signal Corps would obviously need to grow, if it was going to be, in the words of the *New York Times*, "the eyes, ears, and nerves of the Army" on the massive battlefields of this war.[7] When the Army began to search for land for additional Signal Corps training camps to raise qualified cadres of Signal men in sufficient numbers, their investigation led them to Monmouth County, New Jersey. Ownership of the defunct Monmouth Park racetrack land discussed in Chapter 1 had changed hands several times before the Army Signal Corps stumbled upon the site.[8] Notwithstanding the desolation of the largely overgrown and poison ivy infested site in the spring of 1917, it afforded the Army significant advantages. It was roughly 45 miles from Hoboken, which would be a major port of embarkation for troops heading to Europe. It was less than a mile from a train station in the town of Little Silver, which would be incredibly helpful getting men and supplies in and out of the camp. The land had access to both Parkers Creek and Oceanport Creek. And there were good stone roads in the area from the racetrack days.

How did the Army stumble upon this site? Records tell us that Carl F. Hartmann, the Signal Officer of the Eastern Department in New York City, tasked Charles H. Corlett to "go out and find an officer's training camp." Corlett recalled his initial discovery of the Monmouth Park land in a 1955 letter addressed to Sidney S. Davis, Chairman of the Fort Monmouth Traditions Committee. He reported that after examining several other sites, he "finally stumbled on to the old Race Course near Eatontown. I found part of the old steel grandstand with eleven railroad sidings behind it, the old two mile straight away track and two oval race tracks, all badly overgrown with weeds and underbrush." Corlett went on to describe how he arranged a meeting with the owner of the land. "Upon inquiry, I learned that the land belonged to an old man who lived in Eatontown who was very ill (on his death bed in fact), but when he learned my business, he was anxious to see me."[9]

Corlett learned that the current owner, Melvin Van Keuren, had offered to give the land to the Army free of charge during the Spanish–American War. Van Keuren regretfully informed Corlett that he could no longer afford to do so. He offered instead to sell the land for $75,000.[10] Corlett returned to his superior officers to report his findings. Acknowledging the promise of the site, and with the authorization of the Adjutant General of the Army, Hartmann leased 468 acres from Van Keuren, on May 16, 1917, with an option to buy.[11] The *Red Bank Register* newspaper dated June 6, 1917 ran the headline "A Big Farm for Soldiers: Old Monmouth Park Leased by the National Government," and reported that a portion of the land leased by the government had been "farmed for the past four years by Charles Prothero. He will continue to work the farm south of the railroad tracks but all property north of the tracks has been leased by the government. On this property is a 70 acre field of potatoes. The government will recompense Mr. Prothero for this crop."[12]

## Signal Corps Training at Camp Little Silver

The first thirty or so Signal soldiers[13] arrived at Fort Monmouth on June 4, 1917, in two Model T Ford Trucks from Camp Wood on Bedloe's Island in New York Harbor.[14] The next day, a detachment of Depot Company H, Signal Corps, arrived and immediately set out to clear the site. This was a tall order; 129 men would be hospitalized due to their poison ivy in July. The camp was originally named Signal Corps Camp, Little Silver, based merely on its location. General Orders dated June 17, 1917, named then LTC Hartmann the first commander.[15] It is worth spending a bit of time getting to know the fledging fort's first commander, for whom the East Gate, along Oceanport Avenue, would later be named.

Carl Frederick Hartmann was born January 17, 1868 in New York, New York. Before entering military service, he was a practicing lawyer, having graduated from New York University Law School. During the Spanish–American War, he was appointed a "Captain, Signal Officer, United States Volunteers." After service in Cuba, he went to the Philippines and remained there five years. He served at first as the Signal Officer in the Department of Southern Luzon and was later placed in charge of the headquarters over the entire communications system. He went to Alaska in 1905 to work in the Alaska Communication System until 1908, when he attended school at Fort Leavenworth. In 1910, he served along the Mexican border. He was promoted to Major, Regular Army, January 19, 1912, and sent to Fort Omaha, Nebraska, the following year. In 1915 he attended the Army War College in Washington, DC. His next assignment was as Signal Officer of the Eastern Department, with headquarters on Governors Island, New York. There, he was promoted to Lieutenant Colonel, Regular Army, on April 12, 1917, and ordered to establish the Signal Corps Camp, Little Silver.[16] As you can see, it was a very seasoned military man indeed who oversaw the selection of what would become Fort Monmouth.

But back to the camp. Corporal Carl L. Whitehurst was among the first men to arrive at Camp Little Silver. Whitehurst later recalled that the site appeared to be a "jungle of weeds, poison ivy, briars, and underbrush." A 1918 *Dots and Dashes* (the camp newspaper) article called the site "a waste and howling wilderness, thickly overgrown with heavy brush, bramble bushes, chin high rank grass, and swamps."[17] While remnants of the old Monmouth Park racetrack seemed to be everywhere, only one building provided any real shelter. It was there, in that former ticket booth, that Whitehurst and his fellow soldiers, first on the scene, sought shelter as needed while awaiting the delivery of tents. Abandoned railroad cars were put to use in a similar manner. Conditions improved as additional supplies and troops continued to arrive. Within a month, the camp had 25 officers and 451 enlisted men.[18]

Railroads soon brought tents, as well as lumber with which to build barracks. Other early construction included a headquarters building, officers' quarters, transportation sheds, workshops, a library, a post exchange, warehouses, and more. The camp became, essentially, a self-contained town. Unfortunately, most of the lumber used in the earliest construction was green. This would cause problems later as, according to Whitehurst, "By the time the wood was dried out it was winter, and in December there were cracks you could put your finger through. The winter of 1917–1918 was a tough one, and sometimes the snow would pile up on your blankets, coming through the gaps in the boards" in the barracks.[19]

Even as all this infrastructure was being built, training at the camp commenced "amid the fire-gutted ruins of Monmouth Park and Charles Prothero's potatoes."[20] The school course began Sunday night, July 22, 1917, with a curriculum that included physical training, dismounted drill, pitching tents, map reading, camp sanitation, personal hygiene and first aid, cryptography, heliography, semaphore, wig-wag, Army and Signal Corps regulations, and more. Long hikes throughout Monmouth County were common. The plan of instruction called for student officers from each company to act in the capacity of Company Commander, First Sergeant, and Section Chief, so that they would have the benefit of practice in the various grades—a grim reminder of the reality of casualties in the field during wartime.[21]

The training could be intense. John Robertson of Company E, 11th Telegraph Battalion, wrote in a September 11, 1917 letter to his former coworkers that there was always plenty of work to be done. He described the daily routine, as follows: "We are up in the morning at 5:15 and have setting-up exercises until 5:45, breakfast at 6, assemble for drill at 6:45,

A Monmouth Park Racetrack ticket booth welcomed soldiers when they arrived at Camp Little Silver. Pictured in 1917 (top) and 1929 (bottom). (Courtesy US Army Communications-Electronics Command Historical Office, Aberdeen Proving Ground, Maryland)

Camp Little Silver was often informally referred to by the locals as "Camp Monmouth Park" or some variation thereof given the site's former life as the Monmouth Park racetrack. Circa 1917. (Courtesy US Army Communications-Electronics Command Historical Office, Aberdeen Proving Ground, Maryland)

drill until 8:30; one-half hour rest; drill until 10:30, and a fifteen-minute rest, and then another drill until 11:30. Off for the rest of the morning. Mess at noon. The afternoon is taken up with construction work of telephone and telegraph lines."[22]

Another soldier, Edmond J. Brennan, wrote home that month:

> The camp here is in an ideal spot as far as living in tents can make it. Only about a mile and a half from the ocean and forty miles south of New York City. We run up to Sandy Hook quite often from here. All day we have heard warships at target practice off the coast ... The usual day's program is hike practice, signaling, building telegraph and telephone lines and pistol practice. I do not drill much as at present I am driving a motorcycle and side car or truck all the time. The first, second and third battalions got in from El Paso, Texas today and we (the 7th) and the eleventh all go over the blue very soon, they say, I am going up to New York next Sunday to take a little look around. We are half a mile from President Wilson's home and the famous Rumson road where all the millionaires live.[23]

On September 15, 1917, the camp's name changed from Camp Little Silver to Camp Alfred Vail, in honor of the NJ inventor who helped Samuel Morse develop commercial telegraphy.[24] The Governor of New Jersey attended the renaming ceremony.[25] There is an interesting little anecdote connected to this naming. First post commander Carl Hartmann intimated in a 1955 oral history interview conducted by Dr. Thompson, Chief of the Signal Corps Historical Division, that the Chief Signal Officer actually intended the naming of the Camp to honor his good friend Theodore N. Vail, head of the American Telephone and Telegraph Company. In the words of Hartmann, "Recognizing the impropriety of naming the Post for Theodore N. Vail, it requires no stretch of the imagination to figure out why he [General Squier, Chief Signal Officer] came to name it 'Camp Alfred E. Vail'." The impropriety lay in the fact that Theodore Vail was living at the time and was serving as president of AT&T. According to Dr. Thompson, the Signal Corps owed nothing to Alfred Vail, who died a year before the Corps was even established. They did, however, owe a good deal to Theodore Vail. AT&T's peacetime technologies helped facilitate important wartime communications, and, perhaps most importantly, astounding numbers of Vail's employees would volunteer to capably serve the Signal Corps in the Great War.[26]

Regardless of this interesting tidbit, the official name of the post was Camp Alfred Vail, and Camp Vail continued to increase the different training opportunities offered as summer slipped into fall. German language training was one such addition. The importance of producing qualified signal troops in sufficient numbers cannot be overrated—so great was the need that Congress doubled the authorized pay of the Signal Corps private and increased that of the non-commissioned officer.[27]

By October, the first unit left Camp Vail for the War—the Eleventh Telegraph Battalion. Other units quickly followed, to include a Radio Detachment and the 408th Telegraph Battalion in November, and the 52nd Telegraph and 1st Field Signal Battalion in December.[28] One unit, the 1st Telegraph Division, reportedly took their mascot from Camp Vail to France with them. The pup, Smoke, was sadly killed there.[29]

A major impediment to the successful operation of the camp came in September 1918, when the "Spanish flu" epidemic then sweeping the country hit the camp with a fury.[30] Although the Center for Disease Control (CDC) notes that there is still not a universal consensus regarding where the virus originated, it spread worldwide during 1918–1919. In the United States, it was first identified in military personnel in spring 1918.

The CDC estimates that about 500 million people (or one-third of the world's population) became infected with this virus. The number of deaths was estimated to be at least 50 million worldwide with about 675,000 occurring in the United States.[31]

The fifteen men initially afflicted at Camp Alfred Vail were immediately isolated and the units to which they belonged were quarantined. Still, the epidemic spread. By early October, 150 cases were reported, and the entire camp quarantined. A 1918 reports notes, "The Medical Department was called on to exert itself to the limit."[32] The men were forbidden to leave the post, or even to congregate together within its boundaries. Clothing and bedding were thoroughly disinfected; mess equipment was sterilized in boiling water. All work paused. By the end of October, restrictions started to ease and only the hospital units remained under quarantine. Of the total 267 cases treated, many turned into pneumonia. Eleven men died, including Private W. D. Marks of Baltimore, Lieutenant Harold Tierney of Boston, and James P. Harvey of Ventnor, NJ.[33]

Despite battling poison ivy, the hot Jersey summers and snowy winters, and an influenza pandemic, Camp Alfred Vail trained a total of 2,416 enlisted men and 448 officers for war in 1917. The Camp trained 1,083 officers and 9,313 enlisted men in 1918. Between August 1917 and October 1918, American Expeditionary Forces in France received five telegraph battalions, two field signal battalions, one depot battalion, and an aero construction squadron from Camp Alfred Vail. The influx of American troops, trained at places like Camp Little Silver, turned the tide of the war in favor of the Allies. With the signing of the Armistice that concluded the fighting of World War I on November 11, 1918, training activities at Camp Little Silver slowed dramatically.

## The Radio Laboratories

In addition to teaching men how to be Signal Corps soldiers, Camp Alfred Vail was also home to Signal Corps laboratories during the World War I years. The Army wanted to decrease its reliance on private firms like AT&T, though collaborations between Army personnel, private industry, and academic labs would never cease. As future "father of radar" William

The tents shown in the previous photograph soon gave way to extensive wooden construction. Signal Corps Radio Laboratories, Camp Alfred Vail, 1919. (Courtesy US Army Communications-Electronics Command Historical Office, Aberdeen Proving Ground, Maryland)

Blair of the laboratories noted, "The work of these laboratories was, of course, supplemented by commercial companies working under contract."[34] A flurry of new construction dedicated to the new Signal Corps laboratories occurred under the direction of government contractor Charles R. Hedden of Newark beginning in December 1918.[35] Hedden advertised consistently in the local papers for laborers. Orders dated February 23, 1918, named Major L. B. Chambers first commander of the radio labs.

Scientific experimentation by the military therefore took place at the site that would become Fort Monmouth from its very inception. In-house research initially centered on standardization and quality control of vacuum tubes, and various other work related to kinds of radios. Laboratory personnel regularly tested the military suitability of apparatus submitted by private manufacturers like Westinghouse Electric, General Electric, International Radio, General Radio, DeForest Telephone and Telegraph, National Electric, and Marconi Wireless Telegraph. The Camp Vail laboratory buildings were enclosed by barbwire fencing, as the work was generally highly secretive.[36]

Then the laboratory facilities expanded to include one of the first Army airfields, complete with airplane hangars.[37] Flight activity served primarily to experiment with air to ground radios and aerial photography. By May 1918 thirteen planes were on hand (that number would eventually grow to 20). In addition to forty-eight officers and forty-five enlisted men, twelve civilians, principally stenographers, were employed in the laboratories. Within a month, there were close to 100 test flights happening per week. The newly organized 29th Service Company was assigned to duty with the Radio Laboratories, as was the 122nd Aero Squadron. Personnel traveled constantly between Camp Vail and France with models of new equipment for field and battle trials in 1918.[38]

This being the early days of flight, local residents were mystified by the "rasping cough of Liberty motors and the throaty roar of DeHavillands" that "violated the Monmouth atmosphere during 1918," as one early history of the installation put it. It describes how "Residents of Red Bank, Little Silver, Long Branch, and Oceanport were mystified by the labored flight of 'Jennies,' but gradually became accustomed to the aerial circus that flew daily" from the camp.[39]

Flight activities could be fraught with danger, and the camp suffered a loss on August 9, 1918, when Second Lieutenant L. J. Merkel was killed in an airplane accident. His plane went into a tailspin shortly after takeoff and then crashed. Sadly, his brother was visiting and witnessed the accident.[40] Fortunately this was the only serious accident associated with the flight mission of the labs prior to the end of the War. With the November 11 Armistice, flight activity ceased almost immediately. The hangars were empty of planes by mid-December.[41]

As William Blair reflected back on the wartime experience of the laboratories, he felt that "A rather comprehensive program of research and development was carried out by this organization. Many improvements were made to available radio equipment and much new equipment was designed. The Armistice found us near the completion of the program which included a series of radio sets specially designed to meet the requirements of our Army." However, "The time was so short ... that very little of this new equipment had at that time been produced and placed in the hands of troops."[42]

## The Pigeon Service

The use of pigeons by the British and French armies impressed General John J. Pershing, Commander of the American Expeditionary Force. He therefore requested such a service be established in the American Army. This was delayed due to the difficulty in acquiring the birds. The service (consisting of three officers, 118 enlisted men, and a few hundred pigeons) finally arrived in France in February 1918. As many as 572 American birds served in the St. Mihiel offensive; 442 in the Meuse-Argonne offensive. Many pigeons delivered their messages successfully and saved lives, attaining a certain level of celebrity. Cher Ami, for example, was shot through the breast and leg by enemy fire during the Meuse-Argonne offensive but still managed to return to his loft with a message capsule dangling from his wounded leg. The message Cher Ami carried was from a "Lost Battalion" that had been cut off from other American forces. It led to the 194 battalion survivors being saved. Signal Corps Kathy R. Coker and Carol E. Stokes estimate that pigeons like Cher Ami successfully delivered some 95 per cent of the messages entrusted to them.[43]

Pigeons and pigeon lofts, fixed and mobile, were added to the training facilities at Camp Alfred Vail during the World War I years in order to help fill Pershing's need for both pigeoneers and birds for service in France. As such a service had never existed in the US Army, the training had to be built from the ground up. Civilian pigeon racing enthusiasts were enlisted to help.[44] Not just any soldier could be accustomed to pigeon duty. As Ray R. Delhauer, noted "pigeon expert," explained, "the men at breeding and training stations must be carefully trained in their work. Too great care cannot be exercised in the selection of personnel. First of all, they must be men of dependable character. Secondly, they must have a natural liking for birds and animals. It requires at least one year to develop a man to be of any real value the pigeon service."[45]

Following the hectic years of the War, the Signal Corps Pigeon Breeding and Training Service was formally established with a permanent home at Camp Alfred Vail. It existed there until 1957. Birds would be trained to fly two-way missions, and under cover of darkness. Eventually, they could be parachuted into battle; they could be equipped with cameras.

"Hero" birds from Fort Monmouth, credited with saving lives, won awards from multiple nations. These winged warriors are discussed further in forthcoming chapters.

## Economic Impacts

The sudden growth of the camp brought to the area a prosperity which had been absent since the height of Monmouth Park's popularity. The Army locally purchased tremendous quantities of oats, hay, and straw for horses and mules; and cordwood for fires; and produce for the men.[46] Building the camp employed some 300 local men, skilled and unskilled. When there was construction to be done, they worked daily from sun up to sun down Monday through Saturday and were even paid handsomely for overtime work on Sundays. There were non-construction jobs to be had on post as well—even for local women, who worked primarily in clerical roles during this period.[47]

In the local communities, the soldiers themselves proved to be "good spenders," as did those who visited them, and "their relatives, sweethearts, and friends swelled the trade of the storekeepers."[48] To help meet the demand in one town, an addition was built to the Little Silver railroad station for the sale of newspapers, ice cream, tobacco, and soda water.[49] The *Red Bank Register* remarked that the "Little Silver mail route has been extended to take in the camp. This will probably bring an increase in the salary of Arthur Ryerson, the mail carrier."[50] Even jitney drivers "did thriving business, taking relatives of the soldiers to and from the grounds."[51]

## Morale, Welfare, and Recreation

Perhaps unsurprisingly, wartime patriotism and this economic boon meant the local communities were quite welcoming to the soldiers at the post. As one editorial put it, "boys who are preparing to sacrifice their lives for the country should be denied no privilege that will tend to take their minds off the morrow in preparation for what is to come."[52] Many civilians in the area seemed to see the entertainment of the men at the camp as their duty. There was a YMCA operation right on site, staffed in large part by local women. The "Y" arranged musical guests, offered refreshments and games, and even stocked newspapers from across the country, like the *Cincinnati Enquirer,* the *New York Times,* the *New York Sun,* the *Kansas City Star,* and the *Brooklyn Daily Eagle,* so that men stationed at Camp Vail could stay abreast of news from their corner of the country.[53] The varied programming even including English classes geared towards those soldiers for whom English was a second language.[54] There was also a Knights of Columbus "hut" on post, where, "A committee of women from Rumson gained a reputation for showing up at the camp with ice cream," which made the building "a real camp institution whose work is known by every soldier who left Camp Vail."[55] A call for books for the camp was issued by the librarian at the Red Bank Public Library, and local clergy men acted as chaplains.[56] It seemed everyone wanted to contribute what their time, talents, and pocketbooks allowed.

Recreation or canteen houses for soldiers were opened off-post in places like the nearby towns of Red Bank, Long Branch, and Little Silver.[57] The Army & Navy Club in Long Branch was a popular destination, with the *Dots and Dashes* newspaper telling the men, "It's a cinch they know what is good hospitality, and for the benefit, of you birds that, don't know, it's all free—free—get me, kid? Pool tables, checkers, plenty of nice writing paper, and some porch overlooking the ocean, where the cool breezes blow. They have swell chow at almost cost. A big dance every Saturday night, with good music. Give it the once-over; they can't put you in the guard-house for that."[58] In addition to providing entertainment, the locals regularly arranged fundraisers for the camp. Farewell parties were held before units deployed.[59]

Officers and enlisted men were often welcomed into civilian homes for home-cooked meals and fellowship, especially on holidays. A description of one such event tells how:

> Six soldiers from Camp Alfred Vail, Little Silver, were entertained at an old fashioned Christmas dinner yesterday ... by Henry Allen Starks and family ... Dinner was served at 1 o'clock, with turkey as the piece de resistance. In the center of the table was a huge basket of fruit topped with an American flag. At the base of the centerpiece was banked holly, with here and there an American flag. From the center of the table candy ribbons led to the places. Favors in the national tri-colors were distributed and there were also candy filled cornucopias for the guests.[60]

The neighboring towns could even be a bit competitive in their support of the Camp Vail troops, with Mayor John W. Klock of Long Branch declaring, "If is another resort on the continent that done more for its soldier boys than Long Branch, please speak up."[61]

Sports were seen as "clean diversion from strenuous training to increase and keep up the morale of the men;" they also helped to keep the men in good physical condition.[62] There were all manner of sports teams to keep the men busy when they had free time, from football to baseball, basketball to polo, bowling to wrestling, boxing and more. A spirit of friendly competition pervaded these events. As Louis LeVitre wrote home in October 1917, "We had a field day last Saturday and it surely was a great day. There were races of all kinds and an exciting football game. Our battalion made a great showing, making the most points."[63]

When the summer heat got to be too much, "the only relief was found in the waters of the Atlantic," and "whole outfits marched to the ocean … to cool off as best they could" by swimming.[64] Musically minded soldiers formed "The Camp Vail Jazz Band." As the May 1, 1918, edition of the *Dots and Dashes*, noted, "We take our hats off to this little band. They are some band."[65]

Sometimes romance blossomed between soldiers at the post and local women. Carl Whitehurst of North Carolina, introduced earlier in this chapter as one of the first men to arrive at the camp, met and married Estella Irene Ayres of Long Branch. Whitehurst survived the war, and the couple stayed local.[66] In other instances, Mary E. Boyd of Red Bank married Corporal Gerald Dell of Michigan, while Alma Crawford of Freehold married Captain Ralph Bown of Fairport, New York, and Louise Van Allen of West End married James Ross Robertson.[67] Brides sometimes came to the Jersey Shore from across the country to wed their beaus before they deployed, like Edna Avery of Joliet, Illinois. She came to visit Lieutenant Robert Soelke at Camp Alfred Vail, chaperoned by his mother. The young lovers snuck out without Robert's mother knowing, and married in Red Bank.[68] The camp newspaper quipped about marriage in May 1918, "Somebody asked young Stewart, the bugler, who was married a few months ago to a nice girl in Long Branch, whether a man looked at his wife in any different light after they were married than he did before. 'Well,' says he, 'I should say he does! Before they are married he looks at her in the light of half-past ten; afterward he looks at her in the light of five forty-five a.m.'"[69]

On occasion, the locals were more accommodating to the soldiers than the military would have liked. In September 1917, the Army sought the help of local officials in getting saloon keepers, druggists, and even civilians to stop selling alcohol to men in uniform. As an officer of the camp told the press, "I have called the county prosecutors attention to the violations but as yet nothing has been done. There is but one other course for us, and that is to invoke federal aid. Liquor is being sold to our men, in uniform and out of it, some of them having arranged to change to civilian attire in order to obtain their supplies."[70] The local law enforcement and prosecutors were cooperative, and there are records of local men actually being arrested for helping soldier obtain alcohol. As one article reports,

> Abe Bodine ... who was arrested in Long Branch on a charge of having procured liquor for soldiers, was held under $500 bail by United States Commissioner James D. Carton ... Bodine was unable to get bail and was committed to the Mercer county jail to await action by the federal grand jury. Bodine was arrested by Patrolman DeSantis of Long Branch, who declared he saw the man give a uniformed soldier a bottle containing whisky. Bodine is said to be one of several ... who have been making money by buying liquor at saloons and reselling the stuff to soldiers from Camp Vail ...[71]

In another instance,

> Judson McClaskey of Red Bank was held in $500 bail for the federal grand jury by US Commissioner James D. Carton yesterday on a charge of buying beer for soldiers from Camp Vail. McClaskey was arrested at Red Bank Saturday night by two military police who have been conducting an investigation and suspected McClaskey. They approached him on the street and it is said he went to a hotel and purchased some beer for them. He was immediately arrested and locked in a cell at the boro hall. Bail was furnished by Ogden McClaskey, the defendant's brother.[72]

Additional examples abound. It seems the locals could on occasion be a bit TOO accommodating of the soldiers for the Command leadership's liking.

Top: Camp Alfred Vail map, 1919. Bottom: A soldier at Parkers Creek, which is shown in the top of the map, 1917. (Courtesy US Army Communications-Electronics Command Historical Office, Aberdeen Proving Ground, Maryland)

## Great War; Great Success

Despite the profligate ways of some soldiers hellbent on enlisting the locals in acquiring illicit alcohol—by the end of the war Camp Alfred Vail was being called the "best equipped Signal Corps camp ever established anywhere"[73] and "the best equipped radio laboratory in the country,"[74] not to mention "one of the prettiest small camps in the country."[75] One Washington, D.C. man wrote home that "Jersey certainly is a fine state. There are many beautiful homes hereabouts, and the people certainly treat us fine."[76] Soldiers writing in the *Dots and Dashes* declared in one article that "today there are few more attractive and comfortable camps anywhere than Camp Vail" and in another article that "we ... love Camp Vail ... like a girl with her first beau."[77] One soldier/journalist writing in that paper expected that:

> Future years may bring the pomp of power and the pride of place to gladden the hearts of the men who have been in the Training Battalion, at Camp Alfred Vail, Little Silver, N. J., and love of some college Alma Mater may thrill their hearts through the years to come, but the best heritage for those who may live on after this war will be the right to say, "I am a Camp Vail man" ... The espirit of this Camp is indescribable and the courtesy and kindness of the people who have extended so charming hospitality to the men here have been productive of results which will be far reaching always.[78]

Ultimately, the camp trained thousands for war. The radio laboratories worked on pioneering technologies like air to ground radio. The importance of battlefield communications like those facilitated by the camp cannot be overstated. As one doughboy noted, "When the hell begins, Signal contact becomes a man's lifeline. Without it, he is blinded." And as General Pershing himself noted, "... we called upon the highest class professional men in the country during the war and appointed them in the Signal Corps, and through their aid, under the direction of the very efficient Chief Signal Officer, and his own assistants, the Signal Corps in the American Expeditionary Forces established and maintained communication that were better than those of any other army."[79]

The installation was so successful that, while it was supposed to be temporary, it wound up outliving the Great War. The Chief Signal Officer authorized the purchase of Camp Vail in 1919, and the inter-war years would be busy ones.

Field training, Camp Alfred Vail, 1918. The Camp trained thousands of communications specialists for World War I. (Courtesy US Army Communications-Electronics Command Historical Office, Aberdeen Proving Ground, Maryland)

CHAPTER 3

# The Inter-War Years

Radar and Other Research and Development Revelations

As the Great War ended, Camp Alfred Vail was undoubtedly a military success. It had contributed both well-trained signal soldiers and technical know-how (via its laboratories) to the Allied victory. Locals hoped that this would mean the camp, so beneficial to the local economy, would become a permanent fixture, with newspapers hopefully running headlines in early 1919 like "Camp Vail May Be Permanent Post."[1] Remember that while Camp Vail had proved its worth to the military, it was good for the state of New Jersey, too. As discussed in Chapter 2, the sudden growth of the camp brought to the area a prosperity which had been absent since the height of the Monmouth Park Racetrack's popularity. Yet following World War I, the Army was rapidly discharging large numbers of men and slashing budgets. Many of the temporary camps created during the war would have to close.[2]

The first real indications that Camp Alfred Vail might survive as an Army installation despite the post-World War I drawdown occurring Army-wide came in the spring and summer of 1919 when the Chief Signal Officer began requesting that the Adjutant General of the Army move all Signal Corps schools, both officer and enlisted, to Camp Vail. This move would standardize signal communications throughout the Army and consolidate Signal Corps installations. But the Chief Signal Officer asking for Camp Vail to remain open, to host a signal school, did not mean it was a fait accompli. Some camps had to close, as the country was not going to maintain a peacetime Army of millions. Politicians across the country were lobbying to keep posts in their districts open.

An *Asbury Park Evening Press* article in September 1919 warned, "Camp Vail's Future Still Undetermined," explaining, "A temporary signal corps school will be established at Camp Vail, NJ, but final disposition of the camp will not be decided until the permanent army program has been settled."[3] By November 1919, the house military affairs committee would recommend an appropriation of $4,500,000 for the purchase of land for army posts and the completion of buildings thereon, to include $110,000 for the purchase of Camp Alfred Vail.[4]

## The Signal School

The first school commandant of "The Signal Corps School, Camp Alfred Vail, New Jersey" was Colonel George W. Helms. As the first commandant of this institution so central to the history of the post, let's spend a bit of time discussing him. Helms was born in Virginia in 1875 and graduated from the US Military Academy in 1897. As an infantry officer, he served in the Philippines. He was detailed to the Signal Corps in June 1917 and commanded Camp Vail from June 1918 to December 1920. Helms was the first Commanding Officer of the post to be appointed commandant of the newly established Signal Corps School (October 1919–December 1920).[5]

Instruction in the newly consolidated school began October 1, 1919, even as the debate over the purchasing of Camp Vail continued in the Army and in Congress.[6] While in the interest of space this book cannot track the ever-changing curriculum of the school over what would be 50+ years in New Jersey, it is interesting to note that the initial curriculum included an officers' division, subdivided into radio engineering, telegraph engineering, telephone engineering, signal organization, and supply. The enlisted radio specialist course consisted of radio electricity, photography, meteorology, and gas engine and motor vehicle operation. Electrical students were trained as telephone and telegraph electricians. Operator and clerical courses were also offered. The school used the abandoned World War I era airplane hangars as workshops and classrooms since all aerial activity had ceased shortly after the signing of the Armistice. (The old hangars would be used as classrooms throughout World War II.) Courses would be dropped and

The Signal Corps consolidated its training mission at the Signal Corps School at Camp Alfred Vail following World War I. Top: Training at Camp Vail, 1920. Bottom: Instructor personnel, 1936–1937. (Courtesy US Army Communications-Electronics Command Historical Office, Aberdeen Proving Ground, Maryland)

added over the next several decades as military technology evolved and personnel needed to be trained therein; notable additions during the inter-war years that are the focus of this chapter included courses in photography, motion pictures, and meteorology.

Once the Camp Vail land was purchased, work on the Signal School consolidation could really begin in earnest. In May 1920, the Secretary of War asked Congress for $1,500,000 for construction at the Signal School.[7] It would be several years, however, before permanent construction began on post. Still, though the early 1920s were a time of peace, the Signal Corps School expanded as demands for communications training grew. Training of Reserve Officer Training Corps (ROTC) personnel developed into a major function of the school in June 1920, and training began for National Guard and Reserve officers the following year. The school, designed primarily for the training of Army Signal Corps personnel, found itself educating men from several different parts of the Army. The name of the school was officially changed in 1921 to reflect this expanded mission. The new designation, "The Signal School" would be retained until 1935 when it would again become "The Signal Corps School." At other points in the decades to come, the school would function under slightly different names and would train personnel from other branches of the military, and allied countries, when the need arose. Despite these evolutions, the overarching primary mission was always similar in intent: to train men (and later women) to use the latest technologies to win a decisive edge on the battlefield.

## Pershing Visits

The camp's reputation was such that it drew a visit from General John J. Pershing himself. The World War I hero arrived on August 9, 1924. Pershing had been an advocate for keeping the camp open and consolidating signal training there, versus at Camp Leavenworth or Camp Benning, telling Congress:

> ... its present location is most favorable for investigations, and in order to give officers at the school association with men of high professional standing in that line. They would not have that at Camp Benning. There would be nothing of the kind down there. We have the land and the buildings at Camp Vail, and we

have spent about a million dollars there, as I understand it. We have a considerable plant and it seems to me there is every reason why that school should remain at Camp Vail.

When pressed to elaborate, Pershing continued:

At Camp Vail he has the opportunity of mingling with men of high professional standing in the electrical world. Professors from various universities come there to deliver lectures to the officers, who in turn can inspect factories and witness the manufacture of instruments, and all that sort of thing. At Camp Vail they are in close proximity to the factories, so that the manufacturers can very easily send instruments and apparatus in process of manufacture, and the finished product as well, to be tested, and that, to my mind, is a very great advantage, to both the officers and the manufacturers.[8]

Several hundred visitors flocked to the camp to be present as Pershing reviewed the troops. In one of three addresses he made at the camp that day, he reminded troops that their forefathers fought for American independence and that it was up to younger men to uphold the principles of democracy so hard-won in the Revolution.[9]

## Becoming Fort Monmouth

The post's future seemingly secured for the time being, Signal Corps officials and New Jersey politicians continued to try to procure additional funds for construction on post. As the *Asbury Park Evening Press* reported in September 1923:

Although the opening session of the 68th congress is still three months away, Elmer H. Geran, representative from the third New Jersey district, is already busily engaged preparing a number of bills which he intends to introduce. Chief among the measures in which he is interested is a bill for the replacement of the present temporary buildings at Camp Alfred Vail, Oceanport, of a permanent nature and the renaming of the camp "Fort Monmouth." In as much as Camp Vail is one of the points of major importance to the entire army service, Mr. Geran will strongly urge on the floor of the house the passage of this measure. It represents the only large installation devoted entirely to signaling methods and equipment for the entire army. The school and laboratory there are unique in the army and fill a need in the training of army personnel and the development of army equipment which cannot at present be filled with any other agency. Due to the large amount of technical work and investigation carried on along electrical lines, the position of the camp is ideal, situated as it is within easy distance of the largest center for the manufacture and development of communication in the United States, which means in the world. It puts the signal corps in a position to cooperate, as it is doing, directly and intimately with the large civil and industrial institutions engaged in similar work. Another of the advantages of its present situation is the fact that a large portion of recruits for the signal corps are drawn from the northeastern part of the United States ...[10]

The installation was finally granted permanent status and renamed Fort Monmouth in 1925 in honor of the soldiers who fought at the Revolutionary War Battle of Monmouth. Office Memorandum Number 64, Office of the Chief Signal Officer, dated August 6, 1925 stated, "The station now known as Camp Alfred Vail, New Jersey, is being announced in War Department General orders as a permanent military post and will hereafter be designated as 'Fort Monmouth,' New Jersey. Mail to that post will be addressed to Fort Monmouth, Oceanport, New Jersey."

The *Asbury Park Evening Press* reported:

> This action of the war department will be followed by the appropriation of funds to thoroughly modernize the post and to install permanent buildings, it is said, but officers here can not tell how long this will take. It is highly Improbable that any effort will be made to do the work at once, the likely procedure being to spread it out over four or five years, it is believed. Army engineers have drawn up a definite plan for the rebuilding of Camp Vail as Fort Monmouth, which has been sent to Washington for the approval of the war department and the general staff. The post will still continue as school for the signal corps service, it is stated.[11]

The article continued:

> Camp Vail has been rapidly assuming a place in the forefront of army activities ... Here, the men who make possible the vital liaison lines of warfare have been concentrating their activities. The latest field of experiment, which has been attracting much attention throughout the whole shore section due to the use of a huge searchlight, is proving a great success, officers at the post announced this morning. The experiments concern night flying maneuvers and Involve contact between planes and shore stations, as well as landing and direction flying during night flying.[12]

Reading the local reporting, one really senses the locals' pride in "their" army post. The name change was covered by newspapers around the country, though, with the *New York Herald* writing:

> A War Department order changing the name of Camp Alfred Vail, NJ to Fort Monmouth and designating int a permanent military post was made public yesterday ... Fort Monmouth, which is near Monmouth Court House, NJ, where Washington gained an important victory over the British forces under Clinton in 1778, is the home of the army signal school and one of the largest stations in the War Department net. No change in policy will be made at the post. It will remain the principal signal corps experiment and training laboratory of the army for radio, telegraph, and other means of communication. Carrier pigeons which are trained there by the hundreds form the largest collection of homing birds in the country.[13]

The *Washington Post* reported:

> Camp Alfred Vail, NJ, established in the world war on the site of the old Monmouth Park race track, for use of the Signal Corps, and which heretofore had been considered a temporary station, had been declared a permanent post. Its name has been changed to Fort Monmouth, in commemoration of the revolutionary war engagement in that vicinity and in accordance with the military policy of naming all permanent stations "forts."[14]

## Research and Development Laboratories in the 1920s

The laboratories remained among some of the most important facilities at Fort Monmouth during the inter-war years, although sometimes overshadowed by the Signal School during the first half of the Fort's lifetime. (This was due to the sheer number of people associated with the Signal School and the fact that so much of the laboratories' work was highly classified and therefore largely unknown.) The Signal Corps had quickly concluded after World War I that, while private industry and academia made good partners, adequate research facilities for the design and development of Army communications equipment, by the Army for the Army, would continue to be necessary (even if at a reduced scale because of postwar budget restrictions). In William Blair's words, "The war experience had ... clearly demonstrated the possibilities of radio as a means of communication between headquarters in the field as well as the need for continued research and development in order to keep pace with progress in the radio art."[15] Research and development at Fort Monmouth thus continued.

The Signal Corps was anxious to hire the brightest minds the country had to offer, advertising in newspapers across the country. One ad in the *Philadelphia Inquirer* read:

> US SEEKS EXPERTS: Civil Service Commission Announces Vacancies in Radio Laboratories: The United States Civil Service Commission announces open competitive examinations for radio laboratory aid, $900 to $1400 a year; radio labratorian, $1200 to $1700 a year, and junior radio engineer, $1400 to $2000 a year. The examinations will be held throughout the country on December 5. They are to fill vacancies in the Signal Corps, Camp Alfred Vail. NJ, at the entrance salaries named above, plus the increase of $20 a month, and vacancies in positions requiring similar qualifications. The duties are to assist in the development, design and construction of practical and special radio apparatus; to assist in advanced technical work in radio research; to analyze the data accruing from observations of the operation of various radio apparatus and installations; to perform engineering calculations and other related work. Competitors will be rated on the subjects of general physics, mathematics through calculus, practical questions on radio engineering, and education, training and experience.[16]

Among the most important developments of the inter-war period were the SCR-136 ground to ground radio and SCR-134, ground to air radio. These were the first extended range voice radios put into military production.[17] Other projects of this era included the SCR-131, a light and portable unit designed for infantry division and battalion telegraph with a five-mile range to limit possible enemy interception; the SCR-161 for artillery nets; the SCR-162 for contact between coast artillery boats and shore control points; and the SCR-132, a one hundred-mile telephone transmitter with an eighty foot portable, collapsible mast. Other experimentation was performed on items such as tube testers, crystal controller oscillators, unidirectional receivers, and non-radiating phantom antennas. The main thrust with all of these innovations was to ensure that military personnel could communicate with each other as needed, and "see" and react to threats, as quickly as possible on land, air, and sea.

While the military's need for telephone and telegraph communications might seem obvious, the Fort Monmouth laboratories' mission would include work on meteorological equipment. Because, as Signal Corps officer William Gardner Reed wrote in 1922, "Since very early times the influence of the weather on military operations has been of great importance, and success or failure has many times been the direct result of weather conditions."[18] The first radio-equipped weather balloon was launched at Fort Monmouth by 1930. This represented the first major development in the application of electronics to the study of weather and of conditions in the upper atmosphere, as well as an expansion in the function of the laboratory which, prior to 1929, had been primarily to design and test radio sets and some field wire equipment.[19] Consolidation of the five separate laboratory facilities of the Signal Corps was planned that year. The Signal Corps Electrical Laboratory, the Signal Corps Meteorological Laboratory, and the Signal Corps Laboratory at the Bureau of Standards (all in Washington, DC) moved to Fort Monmouth in the interest of "economy and efficiency;" in a manner not dissimilar to the way in which signal training had been consolidated at the post just a few years prior. Conjointly, these laboratories became known as the "Signal Corps Laboratories." The Subaqueous Sound Ranging Laboratory transferred to Fort Monmouth from Fort H. G. Wright, New York, in 1930. (The Signal Corps Aircraft Radio Laboratory at Wright Field in Dayton, Ohio had also been considered for consolidation, but the Signal Corps Aviation Section was abolished.) The Aircraft Radio Laboratory and the Photographic Laboratory at Camp Humphreys became the only research organizations not located at Fort Monmouth. These consolidations represented the first time the personnel and facilities needed to handle almost any Signal Corps problem could be found in one location. The Signal Corps Laboratories employed five commissioned officers, twelve enlisted men, and fifty-three civilians as of June 30, 1930.

Nine crowded wooden buildings constructed in 1918 continued to house the research and development facilities as the 1930s approached. As a result of constant pressure leadership at Fort Monmouth, a $220,000 appropriation was received for construction of a permanent, fireproof laboratory building and shops in 1934. This structure was built under contract. It was scheduled for completion in November 1934, but was not actually completed and accepted until March 1935. It was named Squier Laboratory in honor of Major General George O. Squier, the Army's Chief Signal Officer, 1917–1923.

Much of the communications equipment used by American forces during World War II was designed and developed at Fort Monmouth during the 1930s. The laboratories completed six field radio sets; readied several artillery pack sets for tests; and fielded the SCR-197, a new Air Corps mobile transmitter. The SCR-300 handheld radio (one of the radios often referred to as a "walkie-talkie") was perhaps the best-known development of the period. It has been called the first major development in the miniaturization of radio equipment.[20]

In addition, switchboards, field wire, and radio receivers were developed. One of the most important pieces of equipment developed during this time was radar, which is actually an acronym for Radio Detection and Ranging. The man who holds the patent for the first US aircraft detection system is Fort Monmouth's own William R. Blair, who had been appointed director of the Signal Corps Laboratories in 1930 as a

Colonel William R. Blair, Director of the Signal Corps laboratories and the "father of US aircraft detection radar." (Courtesy US Army Communications-Electronics Command Historical Office, Aberdeen Proving Ground, Maryland)

Major and who served in that position until illness forced his retirement as a Colonel in 1938.

## Blair and America's First Aircraft Detections Radar

Blair was born in Ireland and immigrated to the United States with his parents at the age of nine. He graduated from the University of Chicago with a Ph.D. in Mathematics and Physics in 1906, and joined the Army in 1917. During World War I, Blair served in France and was in charge of the Meteorological Section, Signal Corps. Immediately following the war, he served as a member of the technical subcommittee of the Aeronautical Committee at the Peace Conference.

In 1919, Blair was named head of the Meteorological Section of the Signal Corps, in Washington, DC, and in 1920 he was appointed Major of the Signal Corps in the Regular Army. He graduated from the Signal School at Camp Alfred Vail in 1923 and from the Command and General Staff School at Fort Leavenworth in 1926.

Blair took charge of the Engineering and Research Division of the Office of the Chief Signal Officer that same year. He was promoted to Lieutenant Colonel in 1934, and became director of the Signal Corps laboratories at Fort Monmouth in 1930 until his promotion to full Colonel and retirement in 1938.

In the latter part of the 1920s, governments around the world, multiple different parts of the US military, **and** private industry were in various ways contemplating how one might detect enemy aircraft or ships before seeing them. Blair was unimpressed with experiments in sound detection and felt detection by radio waves was the most promising possibility (hence the word radar originating as an acronym for radio detection and ranging). Locating objects with radio waves was easier said than done. Intensely secret work occurred under the direction of the laboratories at Fort Monmouth (while concurrent work was being done elsewhere by other teams at home and abroad). As news began to leak out about what might be happening at Fort Monmouth, newspapers in 1935 ran headlines like "Army's Mystery Ray," "Mystery Ray Finds Ships at Sea in Dark," "Mystery Ray 'Sees' Enemy at 50 Miles," "Army's New Mystery

Eye Sights Ships Off Highlands."[21] The Army did all it could to keep the details of what they were working on under wraps, with former Fort Monmouth employee William H. Baumgartner recalling in 2002, "They made it super secret. When they talked about it, they pulled down the shades and everything. Even if you **mention** the word it was, by golly, a violation."[22]

Before the United States entered World War II, mass production of not one but two radar sets had begun. The SCR-268 was designed to direct searchlight beams upon aircraft while the SCR-270 was a mobile long-range aircraft detector or early warning set. Newspapers across the country explained to Americans in the summer of 1941 that "Creation of a chain of radio detector stations to warn of the approach of hostile bombers to shores of the United States and overseas bases was ordered ... by the War Department. A call was issued for ... technical experts to man the stations, to be equipped with a secret radiobeam device." Press releases noted that the device "was based on the same principle used by the British in their recently perfected 'radio locators'," but that "The American detector device ... was developed 'entirely independently' by the Signal Corps radio engineers at Fort Monmouth, NJ, over a period of six years. Experts who volunteer to operate the instruments will be stationed at Fort Monmouth for training before they are assigned to operations posts."[23]

In fact, an SCR-270 on Oahu detected the approach of Japanese aircraft on the morning of December 7, 1941. Locating and tracking targets by radio echoes is commonly regarded as one of the most important contributing factors to the Allied victory in WWII.[24] We'll discuss this further in Chapter 4.

The radar invention's top-secret security classification restricted Blair from applying for a patent until June of 1945. He was officially credited with the invention of US aircraft detection radar in 1957, when on August 20 he received Patent No. 2,803,819, entitled "Object Locating System." Military uses of adaptations of Blair's invention have included the detection of aircraft, the direction of anti-aircraft fire, detection and location of ships, and air and ocean navigation. Commercial uses have included aircraft and ship navigation and flight control.[25]

## The Hindenburg Connection

Interestingly, the Signal Corps' radar work included tracking blimps out of the Naval Air Station at Lakehurst, roughly thirty miles away. Scientists at Fort Monmouth had been working diligently on their new radar systems around the time the famed Hindenburg passenger airship exploded into flames on May 6, 1937—but they were not conducting any experiments at precisely the time the Hindenburg met its sad fate. Radar pioneer John J. "Jack" Slattery recalled in a 2001 oral history interview:

> The von Hindenburg, we had expected to track this marvelous target if it got anywhere near us. [But] a storm came up and drenched our equipment. Now, our equipment ... was haywire. It operated on batteries, antennas were held together with wire, all the insulators for the antenna were dry redwood. The storm put us completely out of business. We couldn't have tracked a fly if he was on the transmitter. And, when we heard on the radio what had happened at Lakehurst, when it blew up, the engineer there with me, Jack Hessel, we just ... said "thank god that this rainstorm came!" Because if we tracked them, and then this happened, we felt that we'd wind up in jail or something![26]

No one wound up in jail, and the important radar work continued.

## Camp Evans

Radar testing had been taking place as a number of locations throughout Monmouth County, to include Atlantic Highlands and Sandy Hook. By the fall of 1941, the Signal Corps announced that it would close its temporary radar laboratory on Sandy Hook (which some feared was susceptible to attack or espionage from U-boats), and expand substantially at a new, more easily secured property further inland. The site purchased in Wall Township would at various times be called the Signal Corps Radar Laboratory, Camp Evans, the Evans Signal Laboratory, the Evans Area, and other variations thereof. The "Evans" designation honored the late lieutenant colonel Paul Wesley Evans, a Signal Corps veteran of World War I who died in 1936 from complications due to malaria contracted while on duty in the Panama Canal Zone.

The land had an interesting history, having been used in the early 1900s as a Marconi wireless telegraph station. It then served as a Navy/RCA communications laboratory during World War I, a temporary headquarters for the Ku Klux Klan, and a private college prior to the Army's arrival on the eve of World War II. The Army would use the Marconi buildings and build dozens of additional structures at the site.[27]

## "The Most Amusing Anomaly of the Whole Defense Program"

Even as the laboratories were working on top secret, highly technical projects like radar, the Signal Corps Pigeon Breeding and Training Section formalized at Fort Monmouth following World War I flourished. In 1925, the section had a breeding base with seventy-five pairs of breeders, two flying lofts with one hundred birds for training and maneuvers, and one stationary loft with thirty long-distance flyers. Available facilities permitted the breeding of a maximum of 300 birds per season. That number was banded and held available to fill requisitions from the eighteen lofts scattered throughout the United States and its possessions. Signal School maneuvers and ROTC courses used the birds for instruction. The Officers' Division featured twelve hours of pigeon instruction. In an astounding feat, Fort Monmouth's pigeon handlers successfully bred and trained birds capable of flying under the cover of darkness in 1928.

Despite their serious work, the birds tended to bear the most whimsical of names. Chapter 2 introduced the famed hero of World War I, Cher Ami (who coincidentally, you can visit at the Smithsonian National Museum of American History today). A notable bird of the inter-war years was "Molly Pitcher," named after the legendary heroine of the Revolutionary War Battle of Monmouth (for which Fort Monmouth was named). In the summer of 1930, this renowned flyer was participating in a race in Chattanooga, Tennessee. Somewhere on the 600 mile trek home, she was attacked by a hawk. Her worried handler, Thomas Ross, fretted for three days before receiving a call from nearby Camp Dix that she had landed there. Molly had a big hole in her back, and her weight was down to six ounces from fourteen. As Ross told the press, "She's just a pitiful little handful of skin and bones and she hasn't got many feathers left but she's coming along. She'll fly again." And she did.[28]

On the eve of World War II in September 1941, *Cosmopolitan* magazine ran an exposé on the birds of Fort Monmouth, observing:

> At Fort Monmouth, NJ, US Army Signal Corps base, you can see the most amusing anomaly of the whole defense program. The place fairly reeks with the newest scientific developments in communications. There is every device

The Signal Corps laboratories, along with the school, were a mainstay of the Fort during the inter-war years. While the labs focused on advanced technologies like radio communications and even early radars, the base was also home to what might seem a more antiquated means of communication—the Signal Corps Breeding and Training Center. Top: Laboratory buildings, 1924. Bottom: Lofts, Signal Corps Breeding and Training Center, undated. (Courtesy US Army Communications-Electronics Command Historical Office, Aberdeen Proving Ground, Maryland)

known to 20th century telephony and telegraphy. Yet right in the middle of them all are 2,000 trained homing pigeons—a form of military communications dating back to 43 NC! Despite all our technical progress in wireless telegraphy, the pigeon still remains the quickest method of getting a long message over a relatively short distance. He has many other attributes, too, and when other forms of communication fail, the homing pigeon in the ace in the hole.

The article concludes:

Nobody has yet been able to explain satisfactorily how they do it, but the fact remains that they succeed with almost supernatural regularity. That is why the men at Fort Monmouth, with all their technical genius, have not been able to displace that military messenger who is 20 centuries old: the racing homer![29]

Little did the avian innovators know that this work with the birds would be put to good use when the nation all too soon entered World War II.

## Signal Corps Board

While the school and the laboratories were by far the largest operations on post in the inter-war years, there were other tenants as well—for example, the Signal Corps Board was established at Fort Monmouth in June 1924. This followed a suggestion to the Chief Signal Officer by John E. Hemphill, the fifth Commanding Officer of Camp Vail. Hemphill wrote:

The need for a board of Signal Corps officers to be continuously assembled at a center of Signal Corps activities for the consideration of problems of organization, equipment and tactical and technical procedure has long been recognized. Preferably such a board should consist of officers of considerable rank and length of service in the Signal Corps who would be competent to pass on such equations and would also be able to devote their entire time to the duties of such a board. Due to the shortage of personnel it does not appear that it will be practicable to detail such a board in the near future. The best present arrangement would seem to be a board at Camp Vail consisting of the officers at this post who are immediately connected with the administration and supervision of matters relating to general Signal Corps training. Detailed studies, experimental work, or field tests could be delegated from time to time by this board, with the approval of the Commanding Officer, to the proper subordinates at Camp Vail. It is therefore recommended that a permanent Signal Corps Board be constituted at Camp Alfred Vail to act on such matters as may be referred to it by the Chief Signal Officer.

Army Regulation 105–10 (June 2, 1924) directed the establishment of such a board. Over the years, typical cases considered by the board included the Tables of Organization, Allowances and Equipment, Efficiency Reports, Signal Corps Organizations, and Signal Corps transportation needs. As one man who served the Signal Corps Board in the 1940s recalled, "It was an organization, kind of like the Bell Laboratories used to be, we checked equipment, tested equipment, cable, radios, switchboards, telephones, and that was my assignment there."[30] Fort Monmouth was undeniably becoming the "Home of the Signal Corps."

## Permanent Construction on Post

The "Home of the Signal Corps" needed desperately to augment the temporary World War I wooden construction and tents that dominated the post's infrastructure. Signal Corps men and New Jersey politicians alike had been hoping for a windfall to accomplish permanent construction on post since the early 1920s, but military budgets in the inter-war years were tight. While some money had been found for a bit of critical construction related to the labs, the actual permanent construction program was not approved until 1926[31] and did not begin until 1927 during the command of George E. Kumpe. Four red brick barracks were completed in August 1927 around what is now known as Barker Circle. The construction program continued through 1936, and included: a permanent post hospital, different quarters for officers of varying rank and circumstances, a post theater, a blacksmith shop, an incinerator, a bakery, warehouses, band barracks, utility shops, a fire station, a guardhouse, and expansive laboratory facilities. The quarters of the commanding officer were the last of the quarters to be completed. Arthur S. Cowan, then the eighth commanding officer of Fort Monmouth, first occupied them. The last of the permanent pre-war construction altogether was the headquarters building adjacent to the parade grounds, known as Russel Hall.[32] Much of this construction would be done using local labor, and helped to keep many Monmouth County families afloat during the lean years of the Great Depression—further strengthening the mutually symbiotic relationship that had existed between Fort Monmouth and the local communities since the post's earliest days.

## Citizens' Military Training Camp

Those who drive through Fort Monmouth today can still see much of the permanent construction from the late 1920s and 1930s, much of it gracefully adapted to new civilian uses. Gone without a trace, though, is the Citizens' Military Training Camp (CMTC). This is an oft-forgotten but fascinating piece of Fort Monmouth history during the inter-war years. Authorized by the US National Defense Act of 1920, the CMTC provided young volunteers with four weeks of military training in summer camps each year from 1921 until 1941. Approximately 30,000 trainees participated each year. Those who voluntarily completed four summers of CMTC training became eligible for commissions in the Army Reserve. The CMTC camps differed from National Guard and Reserve training in that the program allowed male citizens to obtain basic military training without an obligation to call-up for active duty.[33]

Charles P. Summerall, commander of the Second Corps Area at Governors Island, NY, discussed the program in 1926, stating, "If a lad attends the complete courses at the CMTC he is of value to the country in time of emergency, for he is then in a position to impart the knowledge he has learned to those who have not taken the courses."

Summerall continued, "But the great feature of the camps is that it gives a lad an opportunity to spend a whole month, free of all expense to himself, out in the open, where he is enjoying plenty of athletic exercise and open air sports, and where he is under the supervision of men who have his best interests at heart."

First-year campers generally received basic military training at the camp in Plattsburg, NY. Students who continued with their courses in subsequent summers could train with engineers, field artillery, coast artillery, cavalry, or the Signal Corps at Fort Monmouth.

Prior to acceptance, the Signal Corps required applicants to demonstrate "a sufficient amount of experience in electrical subjects, either through his school or business training, to ensure that he will have the proper skill and knowledge to satisfactorily fulfill the duties of a signal officer in time of war." Once accepted, trainees learned various means of electrical communication, especially telephone and radio. Instructors taught the reserve officer candidates to install small telephone centrals and to construct field wire lines, as well as to operate the different kinds of army radio sets and message centers. Campers even conducted experimental work in the Signal Corps laboratories.

The trainees dabbled in more traditional signal fields as well. The CMTC at Fort Monmouth schooled trainees in the use of pigeons. The birds came in particularly handy during mock battle maneuvers in late August 1928, when one message sent back to Signal Corps headquarters read, "We are starving, 15 of us without lunch." The *New York Times*, which carried that story, reported that a "chow wagon" was sent to the detachment within 15 minutes.[34]

Camp consisted of more than schooling and maneuvers, though. Plenty of time for recreation existed, as Summerall promised. The day's work ended by mid-afternoon, and tennis, track meets, volleyball, basketball, and boxing were all part of a day's routine. A baseball league was formed in 1927, and the best player on the team won an autographed Babe Ruth bat! Camp chaplain Reverend Clifford L. Miller organized an eleven piece orchestra that year, with plans to play at an end-of-camp dance. Visiting concert parties and movies provided additional on-post entertainment.

The Camp even provided transportation for trainees who wished to spend their free time at the local beaches. Other off-site haunts included the Rumson Country Club, where trainees took part in polo matches, and Red Bank, where they watched airplane races. Students thus clamored to attend the camp at Fort Monmouth. One, Robert J. Boylan, was so anxious to come here that he paid his own railroad fare from his hometown of East St. Louis, Illinois!

Vice-President of the US Charles D. Dawes (1925–1929) declared, "The CMTC present to the youth of this country today an opportunity which should be seized by every young man. These camps teach the advantages and responsibilities of citizenship. They develop students mentally, morally, and physically. They are an asset to the nation." An asset, yes; but the CMTC were disestablished in 1941 as the nation began to mobilize for the possibility of war.[35] Few recall their existence today.

## Fort Fraternization

The locals, who had welcomed the soldiers of Camp Vail from the moment the Signal Corps arrived during World War I, remained as inviting and accommodating as ever to the men stationed at Fort Monmouth in the inter-war years. "Hardly a day passes," noted one general in 1941, without an "event or entertainment given off post by civilians for the soldiers. Large, sandy beaches have been set aside for their exclusive use at no charge. Theaters, baseball parks where night baseball is played, boxing arenas, and other resort centers which dot the coast during the summer have, in most cases, reduced their admission charges for service men ..." Another general noted, "We found that when the spring and summer set in, the first place a lot of young fellows head is the beach."[36] Symbiotically, the post continued to employ local civilians, and those stationed at the post continued to boost the economy by spending money in the surrounding communities.

Love continued to blossom between men stationed on post and local women. In one humorous anecdote from the summer of 1941, seventeen-year-old Jeanne Bohee of Asbury Park and eighteen-year-old Alice Jarcia of Bradley Beach donned military uniforms and snuck on to the installation to surprise their beaus! The *Herald Tribune* reported,

> Two girls dressed in rented Army uniforms were taken into custody ... last night by amazed military policemen who found them wandering around the parade grounds looking for their boyfriends ...
>
> In the post guard house the girls explained between titters that they had invaded the fort for a lark. Their swains, privates in the signal school here, were confined to their quarters, they said, and unable to leave the post to keep a date. The girls decided that if the soldiers couldn't go to them, they would go to the soldiers ...
>
> Military police called in the police of Long Branch, NJ, who took the girls home. Lieutenant Thomas Marks of the Long Branch police said that no charge was placed against them. The girls were not available today when reporters called at their homes. It was said they had gone to the beach for the afternoon.[37]

Security was about to get a lot tighter on post as World War II loomed.

## War Looms; Fort Monmouth Ready

Though the Army overall assumed a peacetime posture in the inter-war years, it invested in the growth of Camp Alfred Vail by purchasing the site that it had originally only leased, designating it a permanent installation, and naming it Fort Monmouth in honor of the soldiers of the American Revolution who had fought and died at the Battle of Monmouth. The Signal Corps' educational missions across the country largely consolidated on post at the Signal School, which increasingly grew the scope of its teaching missions by expanding what it taught and to whom. The Fort's laboratories worked both independently and in coordination with private industry and academia to innovate, adapt, and perfect technologies—like radar—that would help assure the Allied victory in the next global cataclysm.

CHAPTER 4

# "Should They Fail, Expect Plenty of Hell"

Training and Equipment Critical to Winning The Good War

Many hoped that the "Great War," as World War I was initially called, would be the war to end all wars. The Treaty of Versailles, though, rather than establish a lasting peace, unwittingly created many of the conditions which led to World War II. Strongmen like Adolf Hitler, Joseph Stalin, and Benito Mussolini rose to power, hellbent on achieving their expansionist goals at all costs. The Great Depression exacerbated global tensions.

Camp Alfred Vail had outlasted World War I and, in the postwar years, continued both its training and research and development missions. The Signal School ensured a cadre of trained communications specialists existed in the military, and research and development laboratory innovations of the inter-war years would ensure the "Yanks" were equipped to make a decisive difference in the global cataclysm which would, against all odds, erupt and be even more brutal than that which had rocked the world just a few decades prior. The Signal Corps would be more important in World War II than ever before. As one general in the field noted, "The Chief of Staff and myself have limited knowledge of Signal equipment. When Signal communications functions properly, as we expect they will, expect no praise. But should they fail, expect plenty of hell." The troops trained at, and technologies sourced out of, Fort Monmouth would not fail, and would in fact contribute greatly to the Allied victory in what writer Studs Terkel called "the Good War."

## Training Ramps Up

President Roosevelt's proclamation of a state of "limited emergency" on September 8, 1939 following the outbreak of war in Europe had significantly impacted Fort Monmouth even prior to the attack on Pearl Harbor. The Army was immediately authorized additional personnel, and from that point forward the number of men (and soon women) training at Fort Monmouth increased exponentially.

The Signal Corps School curriculum, both officer and enlisted courses, changed to accommodate the increased enrollment. One year following the "limited emergency" proclamation, Congress passed the Selective Training and Service Act providing for one year of compulsory military training. The President simultaneously called the National Guard into Federal service, and the Army increased in size yet again. With the passage of the Selective Service Act, the Chief Signal Officer ordered a Replacement Training Center at Fort Monmouth where enlisted personnel could receive their one year of training. The Signal Corps Replacement Center opened in January 1941. Capacity was initially set at 5,000 men. By December, however, the capacity increased to 7,000 and the one-year training period shrunk to thirteen weeks to address the speed with which men would be needed.

Fort Monmouth's other wartime training focused on officer candidates. The Officer Candidate Department activated within the Signal Corps School on June 2, 1941. The first class commenced July 3, 1941. That class comprised 490 students chosen from warrant officers and enlisted soldiers based on leadership, communications knowledge, and prior service. A total of 335 newly commissioned second lieutenants graduated after three months' training. Subsequent classes averaged about 250 men, but gradually grew to 1,000 men per class.

Tens of thousands of soldiers would ultimately train at Fort Monmouth during the War years at the Signal School, Officers Candidate School, and Replacement Training Center.

## Fort Monmouth Radar Warned of Pearl Harbor Attack

As discussed in Chapter 3, much of the Army's early radar work occurred at Fort Monmouth. Few realize today that radar successfully detected the incoming Japanese planes at Pearl Harbor. Those planes, as Helen Phillips, then Museum Director and Signal Corps Historian recalled in 1967, "catapulted the United States into World War II in a cascade of fire and broken ships. Never since the Revolution had America been so close to ignominious military defeat. Never had our sovereignty been so jeopardized."[1]

Could anything have stopped this infamous attack, and the United States' subsequent entry into the War? Dr. Harold Zahl, known as the "grand old man" of Fort Monmouth research and development, explained in his book *Radar Spelled Backwards* that the SCR-270B radar "gave 52 minutes of unheeded warning as the Japanese approached Pearl Harbor on the 7th of December, 1941." Dr. Zahl had helped to design the vacuum tube components for early warning radars at Fort Monmouth.

Another member of the Fort Monmouth radar team, John Joseph "Jack" Slattery, recalled in a 2001 interview that:

> The 270 (radar project) started at the behest of ... General Hap Arnold, then Chief of the Army Air Corps, when he saw the demonstration (of the SCR-268 radar at Fort Monmouth in) May of '37. (He) asked could the Signal Corps provide him with one of these gadgets that would give his airplanes about a hundred miles warning of an incoming raid to provide them with the opportunity of starting their planes and getting up and attacking the raiders before they got to the target. Well, that said, immediacy was a requirement, far more than the searchlight SCR-268. So, that was set up as a separate group, and they built the SCR-270, first A, and then B, an improved model. And that was the radar that functioned on the Pearl Harbor raid.[2]

In fact, six of the SCR-270B arrived in Hawaii on August 1, 1941. They began testing and operation shortly thereafter. According to one official radar history, General Officers were "highly pleased with the proceedings of radio direction finding sets and the personnel associated with the information centers." Lieutenant Colonel C. A. Powell, Department Signal Officer, stated in a November 11, 1941 memorandum that "We have had very little trouble with the operation of these sets."

So, Pearl Harbor had radar sets capable of detecting the incoming enemy attacks on December 7, 1941. Officers expressed confidence in the capability of that equipment, and, on that "day that shall live in infamy," the SCR-270B successfully detected the incoming Japanese planes.

What prevented the Army from mobilizing to defend Pearl Harbor in the 52 minutes of warning that the SCR-270B provided?

The published findings of the *Army Pearl Harbor Board* and the *Navy Court of Inquiry* reveal what occurred from an operational standpoint, together with an official judgment on the errors in command. Findings are excerpted and summarized below:

> At the time of the attack, only a few vessels of the Pacific Fleet were fitted with radar. The radar of vessels berthed in a harbor such as Pearl Harbor, partially surrounded by high land, is of limited usefulness at best and does not provide the necessary warning of hostile approach.
>
> The shore-based radar on Oahu was an Army service and entirely under Army control. The original project called for 6 permanent (fixed) and 6 mobile installations. The fixed installations had not been completed by 7 December 1941 and only 3 sets had been shipped to Oahu up to that time. On that day there were in operation 5 mobile sets located in selected positions with equipment in efficient condition, but inadequately manned.
>
> Training of personnel had started on 1 November 1941. The Army Interceptor Command was barely in the first stages of organization by 7 December.
>
> Between 27 November and 7 December 1941 the Air Warning System operated from 0400 to 0700, the basis for these hours being that the critical time of possible attack was considered to be from one hour before sunrise until two hours after sunrise.
>
> Two privates, Joseph Lockard and George Elliott, who had been on duty at the Opana station with an SCR-270B from 0400 to 0700 of 7 December 1941, elected to keep running the set until a truck arrived to take them to breakfast so that Lockard, an experienced operator, could give Elliott some instruction. Two minutes after the official closing time for the set a large echo appeared near the extreme end of the range baseline on the oscilloscope when the antenna was facing north. It indicated a large flight of planes at a distance of 136 miles.
>
> So strong was the signal that Lockard checked over the adjustment on the set to make sure it was not spurious. As the echo moved in to a range of 132 miles, there seemed no doubt. Lockard called the Information Center and reported the fact. The Information Center was staffed at the time only by an enlisted telephone operator and an Air Corps officer who was there for training, with very little experience on radar. Having information that a flight of B-17s was expected at the time, he assumed that these might represent the source of the signals and told Lockard to forget it. Meanwhile Lockard and Elliott continued to track the planes in to twenty miles from Oahu, when they lost them due to the distortion.

At this point, according to Lockard's testimony, "we proceeded to close down the station and go back to Kawailoa for breakfast." The two privates were unaware that tragedy was about to strike. The Japanese bombers attacked Pearl Harbor at 0755. Lockard subsequently received the Distinguished Service Medal and an opportunity to attend Officer Candidate School at Fort Monmouth. He was a first lieutenant in the Signal Corps when he testified before the *Army Pearl Harbor Board* and the *Navy Court of Inquiry*. Elliott, who had trained at Fort Monmouth, would also return to the post after his assignment in Hawaii.

Back at Fort Monmouth that December of 1941, the radar team despaired. Had their product failed the fleet? Had it failed the country? Radar technician John Marchetti granted an interview some fifty years after the bombing in which he was asked, "When you heard about Pearl Harbor … how did you feel when you heard what had happened with the radar when the Japanese were attacking?" He replied, "I was furious. I was furious. The radar that had been built … was completely misused. It could have saved lives at Pearl Harbor; it could have changed the picture totally around. But instead it got us in the war, because that was the very real reason why we joined the war."[3]

Jack Slattery's disappointment, as revealed in his 2001 interview, was as fresh as Marchetti's. Slattery's interviewer asked:

> One of the things that is curious about the attack is that it seems as though the radar worked well, but the people in the equation didn't quite follow through the way that they were supposed to, at least the way it's been reported historically. We were curious as to how did the engineers who actually developed this radar- brought it to working efficiency, got it into production- how did they feel about the attack at Pearl Harbor, particularly when they knew that there were six radars set up to protect the island of Oahu and the fleet?

Mr. Slattery replied:

> Well, not only did we know there were the six radars there, but also we knew that each one of them was working properly. Because, we had taken a group of our younger engineers, with maybe a year or two experience, and trained them on how the SCR-270B should perform. And, they'd all been out there and checked them all. So, we were desperately disappointed. And we did not hear until twelve weeks- twelve days later, and we got a very cryptic call from one of the generals in Washington, who said, "Tell the fellas not to worry, their machine worked." But that's all we had for perhaps another month or two. And then we got plots of the tracking, of the tracking of the radars before seven

o'clock and tracking afterwards. And of course, the radar that remained on the air was a voluntary performance of the crew for training purposes while they were awaiting transportation back to their barracks. And the fact that they wanted to work, for more training, was indicative of the loyalty of the crews on all the radars. They would all have stayed if someone had said to them, "can you stay ten minutes and watch something?"[4]

Dr. Zahl recalled similar distress over what he feared may have been the failure of the radars in Hawaii. He remembered that after a few days, "there was still no word as to how or whether our radar had worked at Pearl Harbor. We knew there were investigations on—and we would just have to wait it out. It was around the fifteenth of December. A small aspirin-chewing group had gathered in my office drinking black coffee ... The phone rang ... It was from Washington ... [they] had just received word about our Hawaiian radar. Our Opana Station at the northern tip of Oahu had worked and given some warning. Elated, we cheered, vigorously shook hands and slapped backs ... then silence, (to) hear further details."[5] The radar personnel at Fort Monmouth could rest assured: their radar had worked. The official Army report as to the efficiency of the SCR-270B stated the following:

> The Aircraft Warning Service on the morning of 7 December 1941 was in operative condition for all practical purposes. It had an information center and five mobile stations. It was sufficiently operative to successfully pick up the Japanese force 132 miles from Oahu ... If the radar system and information center had been fully manned, as it could have been and as it was immediately upon the disaster at Pearl Harbor and thereafter without further physical additions, it could have been successfully operated on December 7th ... The aircraft warning service ... was a fully operating service and did so operate shortly after the attack ...

The *History of the Signal Corps Development of US Army Radar Equipment* reports that these facts cast "quite a favorable light at least on the engineering and technical accomplishments of the Signal Corps laboratories personnel in designing, procuring, and shipping out the SCR-270B." Word soon reached Fort Monmouth that the Army wanted more of these sets. Procurement objectives expanded, and research efforts multiplied. The Pearl Harbor disaster had actually proved the indispensable worth of the Fort Monmouth radars.[6]

## Laboratories Fight the Tech War

Radar was hardly all the Fort Monmouth laboratories would be called upon to contribute to the war effort. Personnel worked frantically to make communications for these troops more transportable, more secure, and more reliable. Developments during this period included early FM backpack radios that provided frontline troops with reliable, static free communications, in addition to near continuous advancements in the radar technologies pioneered at the Fort in the 1930s.

For example, Fort Monmouth laboratory development during this period included the SCR-510. This early FM backpack radio, a pioneer in frequency modulation circuits, provided frontline troops with reliable, static free communications.[7] Multichannel FM radio relay sets (such as the AN/TRC-1) were fielded in the European Theater of Operations as early as 1943. FM radio relay and radar, both products of the labs at Fort Monmouth, are typically rated among the top weapon systems that made a difference in World War II.

As one 1942 magazine article put it:

> Tanks and planes are useless without radios, and it is the Signal Corps which furnishes all their radios ... The Signal Corps is, in other words, the nervous system of the Army in all its branches. An Army without proper communications is a mob in a wilderness, and it is the Signal Corps that keeps our indispensable communications going. In the United States, radio and other electrical devices have reached their greatest use and development, and the Signal Corps is symbolic of an electrical and radio minded nation ... It demonstrates the cardinal and essential necessity to use electricity in warfare; electrons go to war.[8]

A "mob in the wilderness" might not defeat the powerful forces of fascism, but the best trained and equipped military in the world certainly would.

## Winged Warriors

The Pigeon Breeding and Training Center remained a staple of post operations even as the Signal Corps' dependence on technology increased. Earlier chapters explained that pigeon handlers had trained birds to fly at night prior to World War II. They were even trusted to fly two-way missions, after in May 1941 20 birds successfully completed the approximately fourteen mile round trip from Fort Monmouth to Freehold Township in half an hour. At that time, the Pigeon Center at Fort Monmouth had an emergency breeding capacity of 1,000 birds a month. This represented about one fourth of the Army's maximum anticipated requirements. American pigeon enthusiasts consequently volunteered 40,000 of the 54,000 birds the Signal Corps furnished to the Armed Services during World War II.

The Chief Signal Officer briefly relocated the Pigeon Breeding and Training Center from Fort Monmouth to Camp Crowder, Missouri in October 1943, but the Center returned to Fort Monmouth on June 20, 1946 with more than two dozen World War II pigeons, including "Yank," "Julius Caesar," "Pro Patria," "Scoop," and "G.I. Joe."

According to "G.I. Joe's" official biography, he made the most outstanding flight by a homing pigeon during WWII. "G.I. Joe" flew twenty miles in twenty minutes to deliver a message from the British 10th Corps Headquarters that ultimately saved the lives of 1,000 British soldiers. The British 56th Infantry Division had requested air support in breaking the German defensive lines at the village of Colvi Vecchia, Italy.

Pigeon training, Fort Monmouth, 1943. (Courtesy US Army Communications-Electronics Command Historical Office, Aberdeen Proving Ground, Maryland)

They captured the village before the air support arrived. "G.I. Joe" reached the Allied XII Air Support Command planes just before they departed for the bombing, which would have resulted in fratricide or "friendly fire."

Major General Sir Charles Keightley, former Commander of the Fifth British Corps in Italy, presented "G.I. Joe" with the Dickin Medal in a 1946 ceremony at the Tower of London. The Dickin Medal was awarded for gallantry by the People's Dispensary for Sick Animals. It is known as "the Animal's Victoria Cross." "G.I. Joe" was the first non-British animal to receive such recognition. He later received Congressional recognition in the United States.[9]

Many birds like "G.I. Joe" achieved celebrity status. Newspapers across the globe covered their wartime exploits and, post-war, published effusive obituaries when they passed.

## Skirted Soldiers Invade Post

Fort Monmouth had been a largely male dominated space up to this point, though women did work in clerical roles and help to run morale, welfare, and recreation endeavors back to the post's founding. This is in keeping with the fact that, prior to the World Wars, the Army only occasionally used women and in what it considered gender appropriate roles. For example, civilian women, often known as camp followers, cooked and mended for soldiers during the Revolutionary and Civil Wars much as they had done for their men in times of peace. A few women acted as nurses during the American Revolution and continued to do so after the country gained its independence, despite popular concerns about the close contact with males this work required.

Early twentieth century advances in military communications-electronics technology, combined with manpower shortages, provided the Army with opportunities to employ women in less traditionally feminine roles. The Signal Corps consequently used women as telephone operators during World War I (though there is no record of any of the "Hello Girls" passing through Fort Monmouth). The Corps initially recruited bilingual women from commercial telephone companies, but later accepted less experienced applicants to fill the growing demand. According to historian Rebecca

Robbins Raines in *Getting the Message Through: A Branch History of the US Army Signal Corps*, the first unit of female telephone operators to serve with the American Expeditionary Forces arrived in Paris in March 1918. Approximately two hundred female telephone operators ultimately served in operating units in the First, Second, and Third Army Headquarters. The women worked in Paris and dozens of other locations throughout France and England. Nicknamed the "Hello Girls," these women worked long hours, often under combat conditions.

In one instance, the Army forcefully evacuated the Female Telephone Operators Unit of the First Army Headquarters because the women refused to desert their posts even after their building caught fire. The women, after readmittance to the building, restored operations within an hour. They subsequently won a commendation from the Chief Signal Officer of the First Army. Grace Banker, chief operator, even received a Distinguished Service Medal for her wartime service.

World War I Chief Signal Officer Major General George Owen Squier later cited women's "unquestioned superiority" as switchboard operators and their value in freeing men for the fighting front. The Report of the Chief Signal Officer, 1919, declared that "The use of women operators throughout the entire war was decidedly a success ..."

That success notwithstanding, the Signal Corps released its "Hello Girls" soon after the Armistice. Unfortunately, according to the US Army Signal Corps museum, the female operators returned home only to realize that "all Army regulations were worded in the 'male' gender, so the women were denied veteran status. They were considered civilians working for the Army. This perplexed the women because they were required to wear regulation uniforms, they were sworn into service and had to follow all Army regulations." Only decades later, in 1978, did legislation award the operators veteran status. Despite the regrettable lag in official recognition, proponents of the gender integration of the Army during World War II often cited the Signal Corps' successful employment of the "Hello Girls."

Manpower shortages during that conflict again necessitated the use of women. Historian Russell Weigley explains in his book *Eisenhower's Lieutenants* that by mid-1943 the Army was approaching "the limit of the numbers it could remove from the economy without endangering a basic conception of the Allied war effort, that America was to be the industrial arsenal for all the Allied powers." According to Max L. Marshall's *The Story of the US Army Signal Corps*, the Corps, in particular, required an enormous increase in personnel to cope with "the new equipment developments and the mass production of wire and cable, radio radar, and all the increasingly complex components of modern communications-electronics."

The utilization of women offered what historian Karen Kovach has called a "golden opportunity" to solve these labor shortages. So recognizable was the opportunity that Chief of Staff General George Marshall himself told the War Department in November 1941, "I want a women's corps right away, and I don't want any excuses!" The 77th Congress eventually did establish the Women's Army Auxiliary Corps (WAAC) with Public Law 554 on May 14, 1942, which allowed women to serve "with," not "in," the Army. The law passed after much heated debate amongst chauvinistic Congressmen over whether or not "women generals would rush about the country dictating orders to male personnel and telling the commanding officers of posts how to run their business," and "who then will do the cooking, the washing, the mending—the humble home tasks to which every woman has devoted herself?"

The Army thus became the first of the services to enlist women during World War II. Key members of the Signal Corps advocated the use of women early in the conflict. This contrasted with much of the Congress, many military personnel, and a high percentage of the general public, all of whom considered a woman's place to be in the home. The Signal Corps, by 1942, had identified 2,000 jobs suitable for WAACs. Hundreds of auxiliaries soon descended on Fort Monmouth and other Signal Corps posts.

The Commanding Officer of the 15th Signal Training Regiment at Fort Monmouth, Colonel Frank H. Curtis, claimed of these women, "their pedagogical training has given them a well-diversified background … They grasp their work rapidly, and have a keen sense of loyalty. We're glad to have them working with us." This type of praise provided a marked contrast to the ongoing, public "slander campaign" against WAAC morality, which became so widespread that First Lady Eleanor

Roosevelt, Secretary of War Henry L. Stimson, and General Marshall all publicly denounced it.

Despite such obstacles, the WAAC gave way to the Women's Army Corps (WAC) in 1943. This made women a part of the Army as opposed to an auxiliary thereof. Signal Corps Colonel Harry O. Compton outspokenly supported this decision, declaring, "They (women) are particularly adept at work of a highly repetitive nature, requiring light, manual dexterity ... Where men grow tired and bored, women's efficiency remains unimpaired." Even WAC Director Oveta Culp Hobby recognized women's usefulness to the Signal Corps, stating, "From the inauguration of the WAC, the potential usefulness of members in carrying out Signal Corps duties was recognized." The Signal Corps would be the first agency of the Army Service Forces to request Women's Army Corps personnel and utilized a higher percentage of female replacement communicators than any other technical service, except the Chemical Warfare Service. Signal Corps WACs in Europe represented 23.5 percent of all the WAC personnel in the Communications Zone, exclusive of the United Kingdom. This meant approximately one Signal Corps WAC for every 55 Signal Corps men. The other services could claim only one WAC to every 234 men.

More than 150,000 women would serve with the Women's Army Corps during World War II. Around 5,000 served with the Signal Corps. A Signal Corps board meeting convened at Fort Monmouth post war, in 1948, eventually contributed to the Army's decision to retain women in peacetime. The Board deemed women "more adaptable and dexterous than men in the performance of certain specialties." All of the military services eventually lobbied Congress for the continued participation of women in the Armed Forces. The Women's Armed Services Integration Act of 1948, Public Law 625, passed by the 80th Congress, consequently gave women a permanent place in the military services.

It was not until 1978, however, that the Army abolished the WAC and fully integrated women into the Regular Army. The Signal Corps, with its "home" at Fort Monmouth, had championed women for the past sixty years. Its use of civilian female telephone operators during World War I represented one of the ways the Army first cautiously used women in what it considered gender acceptable roles, outside of nurses and camp

"SHOULD THEY FAIL, EXPECT PLENTY OF HELL" • 69

FRIDAY, AUGUST 20, 1943.     Published Weekly

### Skirted GIs: They're In The Army Now
## Post's Women-Soldiers Sworn Into Army Before Two Generals

### WAC Units Here Take Oath During Solemn Ceremony

**Auxiliary Status Ends As Girls Enter Ranks Of Service Corps**

To the strains of "You're in the Army Now" and the cheers of spectators who came to see one of the post's most inspiring spectacles in months, Fort Monmouth's WAACs Wednesday evening, following in the path of thousands of their comrades-in-arms throughout the nation, were sworn in as members of the new Women's Army Corps.

Described by Capt. Florence E. Mead, sprightly commanding officer of the WACs, as "one of the greatest thrills of my life," the ceremony was reviewed by Brig. Gen. George L. Van Deusen, commanding Eastern Signal Corps Training Center; Brig. Gen. William O. Reeder, commandant of the Eastern Signal Corps Schools; Lt. Col. Donald McLean, Executive Officer; Lt. Col. John S. Weeks, Director of Personnel; Major Duncan E. McKinlay, Special Service Officer, and Capt. Roger E. Lawless, aide to General Van Deusen.

**Oath Administered**

The oath of the WAC was administered by Colonel Weeks. Bright orange guidons fluttered strongly in the sweeping breeze that ran down the main parade grounds as the four platoons of WAACs made the long march across the broad, open field. Long, contrasty shadows made by the evening sun set off a moving picture of mechanical precision, turning to a rythm of silent cadence.

Seconds after Capt. Mead and her two lieutenant WAC aides reached the reviewing place from the field, Col. Weeks was at the microphone with raised hand administering the

(Continued on Page 6)

### Wood Radio Unit Transferred To Fort Jurisdiction

**Schools, However, Will Remain With Exeption Of Field Radio Classes**

Effective last Monday, August 16th, the Radio Section of Signal Communications Division at Camp Charles Wood was transferred to the jurisdiction of Fort Monmouth. Under the new set-up, it will be a part of the Code and Traffic Section of the Radio Division of the Eastern Signal Corps Training Center at Fort Monmouth.

The schools will remain at Camp Wood with the exception of Field Radio. Heretofore, field radio was included in the course at this Eastern Signal Corps Unit Training Center, but will now transfer to the main post.

Radio operation, procedure and the combined means of visual communication will be the basis of the course taught at Wood. As previously, the qualifying speed of fifteen words per minute sending and receiving will be maintained. The

Photos by Pfc. C. E. Rosen

They're in the Army now!! Standing smartly in formation, members of Fort Monmouth's detachment of the Women's Auxiliary Army Corps solemnly take the oath of allegiance which changes their status from an auxiliary force to the Women's Army Corps.

Lower photo shows Aux. Helen G. Patterson with her mother, Mrs. Jeannette Patterson, who is a veteran nurse of the Spanish-American War and was a member of the first Army Nurse Corps during the Philippine insurrection in 1899. Her brother, Commander Richard O. Patterson, is on active duty.

### GI Forum Agrees

Excerpt from the *Signal Corps Message* newspaper announcing the transition from the Women's Army Auxiliary Corps to the Women's Army Corps. August 20, 1943. (Courtesy US Army Communications-Electronics Command Historical Office, Aberdeen Proving Ground, Maryland)

followers. Breaking down the barriers that impeded camp followers and even Army nurses, these pioneering women answered the call that would integrate by gender what historians have called the "most prototypically masculine of all social institutions," the United States Army.[10]

## Expansion Required

The number of people working on Fort Monmouth expanded until the infrastructure was maxed to capacity. The post had started expanded even prior to the US entry into World War II, On August 12, 1940, Chief Signal Officer Joseph O. Mauborgne advised then Signal Corps School commandant Dawson Olmstead, "in the event that the reservation at Fort Monmouth is inadequate ... it is desired that you ascertain ... the feasibility of leasing additional ground in the vicinity of Fort Monmouth ..." Olmstead realized that the present area of the post, approximately 440 acres, was indeed insufficient to accommodate the post's burgeoning missions. He thus recommended the purchase of three contiguous areas. The acquisition of these properties, and the filling of a portion of Parker Creek, increased the size of the main post to 637 acres. On this acreage, the Army built encampments wherever it could find vacant space.

The next year, then commander of the aforementioned Replacement Center, George Lane Van Deusen, initiated the purchase of two contiguous properties about two miles west of Fort Monmouth (a site that would come to be known as the Charles Wood Area). This second expansion of the post proved insufficient still. Chapter 3 discussed the consolidation of radar work at Camp Evans in Wall Township, and Fort Monmouth resorted to leasing numerous other properties in Monmouth and Ocean Counties in order to accommodate its wartime missions, sites many have since forgotten were once associated with the defense of our nation.[11]

Perhaps least surprising, in 1941, Olmstead leased the National Guard Encampment at Sea Girt for $1 a year, plus $125 a day for power, gas, and water. The camp, according to Helen Phillips in her 1967 *History of the United States Army Signal Center and School*, "boasted twenty-two mess halls, latrines, a small laundry, an exchange, and concrete tent floors on which to construct the major portion of the temporary buildings." According to Olmstead, "because of its extensive drill ground, target range, and railroad sidings, it is ideally equipped to permit the reception of selectees." Upon completion of three weeks' basic training at Sea Girt, selectees would be transferred to the main Replacement Training Center at Fort Monmouth to continue their specialist training. The War Department named the site "Camp Edison" on April 25 for Charles Edison, governor of New Jersey and son of the inventor, Thomas A. Following the war, the Army returned Camp Edison to the New Jersey National Guard. The land is still used by the National Guard and the New Jersey State Police.

Camp Edison seemed purpose built for Fort Monmouth's overflow needs, but Asbury Park's Convention Hall, a beachfront landmark and beloved New Jersey treasure, was a more interesting choice. Basically it was available, and it was large. Colonel Willard Matheny, Assistant Executive Officer at Fort Monmouth, revealed the Army's plan to lease Convention Hall in August 1942. He explained that the expansion of the Signal Corps at Fort Monmouth had been so rapid that there was no time to build sufficient facilities on post. The Hall would serve as a training facility for officers. Helen Phillips reveals that the Army divided Convention Hall into sixteen classrooms, including two large rooms that each provided seating space for 150 students at map tables. While the classrooms were being partitioned off, instruction was conducted in the Asbury Park YMCA gymnasium. The Signal Corps had also leased that facility. Archival records show that the building, completed in 1929, primarily housed the Eastern Signal Corps School Enlisted Department. Convention Hall and the YMCA were released from military service in November 1943.

The Army also drafted lesser known properties, like Asbury Park's the Marine Grill. Records show that this restaurant served as a mess facility for 325 officers. Many of those officers training at the architecturally brilliant Convention Hall and dining at the elegant Marine Grill lived at the plush Kingsley Arms Hotel and the Santander Apartments, both also in Asbury Park. Soldiers housed there must surely have felt as if they won the lottery! The Marine Grill, Kingsley Arms, and the Santander all retired from military service in November 1943.

The Army Signal Corps later enlisted the 150-room oceanfront Grossman Hotel on Ocean Avenue in Bradley Beach. Called into service in early 1943, Grossman Hotel was used not for housing but for office space for Signal Corps procurement functions. Utilization of this facility consolidated functions previously scattered between Fort Monmouth and its satellite site, Camp Evans. Approximately 400 Army and civilian personnel worked at the Hotel during the War.

Fort Monmouth's list of leased facilities used during World War II also includes the Sea Girt Inn, drafted into service in the spring of 1943. Records show that the facility, once closed by the local authorities for staging "indecent performances," was used for far more serious purposes by the Signal Corps Engineering Laboratories. The *Times* reported that the Inn was used for research and development to benefit the Army Air Force. Air Force Colonel (Ret) Albert C. Trakowski later recalled, "It was an old nightclub that the Signal Corps had rented for the purposes of doing remote experimentation ... Rawinsonde. I did most all the work on developing how to use that equipment ... there at the Sea Girt Inn." According to *Army Air Forces in World War II, Vol. VII: Services Around the World*, the rawinsonde was an adaptation of radar that proved especially helpful to air operations by allowing the measurement of wind velocity and direction at high altitudes without the necessity for optical tracking previously required. Important work for a site once accused of hosting burlesque shows![12]

## Civilian Impacts

Just as civilian properties heeded Uncle Sam's call to serve, the wartime demands on Fort Monmouth created additional job opportunities for civilians from local communities and further afield. As Marge Bramley recalled in a 2000 oral history interview, "My sister and I both worked at AT&T in New York at 195 Broadway as messenger girls. We made fifteen dollars a week and we paid three dollars and fifty cents a week for commuting to the city. We were just surviving. One day my sister took a day off and went over to Fort Monmouth because she had heard they were starting to hire people. When I came home, she said,

'Maggie you're never going to believe it. Guess what? I got a job as a typist, and they're paying me forty-five dollars a week! From fifteen!' Guess what? I took the next day off and did the same thing ... And my mother got a job over there too. All three of us. So that's what we did during the war."[13]

As during World War I, there was again an influx of people who moved to Monmouth County to be near loved ones stationed at the Fort. Joan Austin of Ridgewood, NJ, for example, was a new bride who set up housekeeping in a small apartment near the post. She lived there with her new husband and two of his West Point classmates, as a housing shortage had developed in the area. As *Good Housekeeping* reported:

Military and civilian employees working hand in hand in support of the war effort, 1943. (Courtesy US Army Communications-Electronics Command Historical Office, Aberdeen Proving Ground, Maryland)

Housekeeping occupies most of the day, as Mrs. Thompson, with bridelike fervor, experiments with recipes to tempt the hungry men, rearranges the furniture trying to make more room, keeps order in the closet where his Army laundry sack nudges her neat chintz dress bags and boxes. Twice a week she rolls bandages at the Fort, once a week plays bridge with other officers' wives. Saturday night she steps out, squired by her husband and grateful boarders. She loves dancing. She says the Army is the life for her. She's wedded to it![14]

## POWS during World War II

One interesting story Marge Bramley shares in her oral history interview relates to Italian prisoners of war who arrived at Fort Monmouth to perform housekeeping duties. A lieutenant colonel and some 500 enlisted men became hospital, mess, and repair shop attendants, relieving American soldiers from these duties.[15] Marge recalls,

> We ... had a great number of Italian prisoners. They were all billeted at Fort Monmouth ... They had a whole section where all the Italian prisoners were kept. They worked on the yards and stuff, you know, lawn work, and jobs like that. But on a Sunday the whole place was surrounded by Italian-Americans. The Italians were all billeted on Oceanport Avenue, and they had a big wire fence that ran all around it. The Italian-Americans would come every Sunday. Tons of people from Long Branch, which has always been an Italian settlement, would come, because they knew some of these families. You would see a long line of them; they would bring food and treats. They all knew how to speak Italian. At first, it was like, "Ah, they're out there with the enemy," and then you'd think, "Oh, well they're Italians, they are not really our enemies." It was quite interesting to see how they were soon accepted, and I'm sure some of them married local girls.[16]

Marge's perspective on Italian POWs is a fascinating one. The *Asbury Park Evening Press* seemingly reinforces the smooth experience of these POWs, reporting, "Generous supplies of cigarettes, butter, and soft drinks are the principal reason why prisoners at Fort Monmouth and other New Jersey military facilities have caused comparatively little trouble ... New Jersey prisoners are permitted two packages of cigarettes a day, generous supplies of pure creamery butter, and all the soft drinks they want ... At Fort Monmouth, Italian prisoners have made no attempts to escape and no trouble has developed between the American and the Italians."[17]

And it seems at least one local girl did marry a former prisoner: Judith Gatta of Bangs Avenue in Asbury Park was reported to have sailed to Rome in April of 1946 to wed Lieutenant George Ricci, who she met while he was imprisoned at the Fort.[18]

## Training Adjusts As Needed

Wartime training subsided quickly as the tide of the war turned. Reductions began as early as May 1943 with orders to inactivate the Replacement Training Center. The center produced more than 60,000 Signal Corps specialists during the thirty months of its existence. Then, in August, the capacity of the Officer Candidate School was set at 150. Classes entered at seventeen-week intervals. Enrollments fluctuated thereafter. Most of the functions of the Enlisted Department of the Signal School transferred to Camp Crowder, Missouri in 1945 with the decline in requirements for trained replacements within the Signal Corps (although many of these functions would return in 1946—the Army **does** love its restructuring and reorganizing).[19]

## Cold War on the Horizon

Despite changes to the Fort's educational missions even prior to the War's end, research and development continued throughout Germany's surrender on May 8, 1945 and Japan's on August 15. Innovations of the latter 1940s by Fort Monmouth personnel included the first weather radar, which observed a rainstorm at a distance of 185 miles and tracked it as it passed over the Fort; and the development of synthetic quartz, which freed the military from reliance on foreign imports of the critical mineral. A technique for assembling electronic parts on a printed circuit board pioneered the development and fabrication of miniature circuits for both military and civilian use.[20]

Perhaps the most mystifying development of the immediate post-World War II era came from the Evans Signal Laboratory (aka Camp Evans), which remained a sub-installation of Fort Monmouth despite the end of the war. A true milestone in scientific history occurred on

January 10, 1946 when Signal Corps scientists, under the direction of John J. DeWitt, Jr., used a specially designed radar antenna (called the Diana Tower) to successfully reflect electronic signals off the moon. The project was named in honor of Diana, Greek goddess of the moon.

The Diana antenna focused a beam of high frequency energy at the moon, traveling at the speed of light (186,000 miles per second). Scientists achieved success shortly after moonrise when an audible ping came over the loudspeaker of their receiver, signaling the return of the radio wave just two and a half seconds later. Continuous recordings were made at regular 2.5 second intervals. The Diana experiment proved the feasibility of communicating across vast distances of space, and newspaper reports at the time put the feat into the same category as the development of the atomic bomb. Actor Orson Welles, who inspired fear of a Martian attack on earth in 1938 with his *War of the Worlds* broadcast, announced the Diana feat on the radio over WJZ and the American Broadcasting Co. network. He relayed, "Radar contact has been established with the moon." He quickly added, "Ask the scientists of the Signal Corps if you don't believe me. And I can't blame you if you don't."[21] The Diana radar experiment was a significant milestone in history, but also a mere precursor to the Cold War era innovations to come—as you will see in Chapter 5. As one popular magazine noted, "It once seemed like magic to the people who were not electronics experts," but radar, and other Fort Monmouth innovations, would soon be impacting everyone's daily lives in addition to making a decisive difference on the battlefield.[22]

CHAPTER 5

# "Where the Army Signal Corps Thinks Out Some of the Nation's Crucial Defenses"

## Cold War Battleground

The end of World War II did not necessarily mean that the United States assumed a peacetime footing. It had gained new international importance following this last great conflagration, and plunged right into the Cold War with its begrudging wartime ally, the Soviet Union. Fort Monmouth would be on the front lines of capitalist, democratic America's decades-long struggle for global hegemony with the communist Soviet Union—even employing former German scientist to help win a technical edge. The *New York Times* referred to the post as "where the army signal corps thinks out some of the nation's crucial defenses."[1] This notoriety drew the attention of infamous Senator Joseph McCarthy, who descended upon the post convinced a nest of Soviet spies operated on site. Dozens lost their jobs, as the post commander cooperated with McCarthy. Despite that unfortunate incident, the post's schools continued to train men (and increasing numbers of women) and its laboratories proceeded with pioneering work that would impact the space race, the Korean War, and so much more.

## Korean War Support

Following World War II, the defeated Axis powers withdrew from areas they had occupied. New countries emerged; new borders were drawn.

The Soviet Union hoped to remake the world in its image; United States quickly became committed to containing communism—believing if one country fell to it, its neighbors would also fall, like dominos. Within this context, in 1948, the Korea Peninsula—which had been occupied by the Japanese during the war—was divided between a Soviet-backed government in the north and an American-backed government in the south. War broke out along the 38th parallel on June 25, 1950. On that day, North Korean troops coordinated an attack at several strategic points along the parallel and headed south toward Seoul. The United Nations, formed in 1945 to, hopefully, help ensure a postwar peace better than the League of Nations had done, responded to the attack by adopting a resolution that condemned the invasion as a "breach of the peace." President Harry S. Truman quickly committed American forces to a combined United Nations military effort and named General Douglas MacArthur Commander of the U.N. forces. Fifteen other nations also sent troops under the U.N. command. Truman did not seek a formal declaration of war from Congress, but war raged until 1953. Some five million soldiers and civilians lost their lives, to include close to 37,000 Americans.[2]

At times when the Cold War turned hot, as in Korea, Fort Monmouth's technologies continued to make a decisive difference on the battlefield, as they had in World War I and World War II. This time, though, much of the laboratory work for the troops focused on tweaking technologies innovated during the last war. As Thomas Daniels, who arrived at the Fort in the early 1950s, recalled, "In the Korean War, officially what we were attempting to do was expand the range of the equipment or improve on the effectiveness of it. So we took equipment that was already in existence and we modified it to improve its range and accuracy."[3] Automatic artillery and mortar locating radars, like the AN/TPQ-3 and AN/MPQ-10, proved among the major successes, helping soldiers to detect the source of incoming enemy attacks and to potentially launch counterattacks.[4] *Popular Mechanics* observed, "Infantrymen who have exhausted their vocabularies on mortar fire that pinned them down will welcome a new Army radar that spots enemy mortars by tracing the trajectory of projectiles. Developed by the Signal Corps and Sperry Gyroscope Co., the mortar spotter ... consists of a dish-shaped radar antenna, automatic tracker, remote-control console with radarscopes, a gasoline powered motor-generator and a trailer mount. Backtracking the projectiles, the radar pinpoints the enemy mortar."⁵

Other developments of the period put to use in the war included lightweight field television cameras, pocket radiation detectors, and significantly smaller field radios. These included the PRC-6 and PRC-10 radios to replace bulkier models like the SCR-510. While the SCR-510 was intended to be mounted in a vehicle, the PRC-10 weighed roughly 23 pounds and could be carried in a backpack As the Signal Corps Engineering Laboratories 1951 annual review noted:

> ... an entirely new line of radio sets was becoming available for troops in the field which will provide a large portion of the radio communication facilities in the forward area and, having been designed with stress on miniaturization, weatherproofing, stability, and ruggedness, will meet the stringent demands of severe field usage with a higher degree of satisfactory performance than ever before achieved.⁶

Installing communications lines in Korea, August 1950. (Courtesy US Army Communications-Electronics Command Historical Office, Aberdeen Proving Ground, Maryland.)

The idea was to make radios lighter, more transportable, and sturdier, to push reliable communications further down into the ranks.[7] It bears repeating that many of the innovations discussed in this book were to some degree or another joint projects where multiple Army institutions, academia, and the private sector played a hand. As historian William R. Stevenson noted in his 1966 monograph *Miniaturization and Microminiaturization of Army Communications-Electronics, 1946–1964,* "It would be grossly unfair to the creative electronics industry ... of the United States and to the other military services to say that the Department of the Army was the principal exponent of this [electronics] revolution—but, with more becoming modesty, the Department of the Army can claim that it has supplied much of the revolution's ammunition."[8]

## The Signal Corps Labs Reach into the Sky

The Fort Monmouth laboratories were not solely focused on modernizing existing equipment for battlefield troops during this era. The space race is perhaps one of the more constructive elements of the Cold War, and Fort Monmouth personnel worked furiously on brand new satellite technologies. When the Russians launched Sputnik I in October 1957, the Fort's Deal Test Area some eight miles away from the main post was the first government installation in the United States to detect and record the Russian signals. The Deal Test Area is yet another example of how the post acquired and used lands in addition to main post whenever the need arose. The Army had acquired this particular 208 acre parcel of land in Ocean Township in 1953. It had earlier been used by Western Electric, which acquired the plot in 1919 and began an experimental station for ship to shore radio telephony. The property transferred to Bell Laboratories in 1925. Extensive radio communications and radar experiments occurred there between 1920 and 1953.[9] A *Bell Labs Bulletin* dated January 22, 1953 tells us that "Many other significant advances were born at Deal, including the development of the short wave transmitter for the first transatlantic radio-telephone service. During WWII, a part of the Laboratories' work on radar research and development was carried on at Deal." The Bell Laboratory work at Deal merged with the Bell Laboratory work being carried out at Holmdel in 1953. US government activity at the Deal Test Site began in September. The US Army Corps of Engineers, District of New York leased the property; however the using agency was the Signal Corps headquartered at Fort Monmouth. Fast forward to October 1957, and a total of 273 orbits of Sputnik I were observed and recorded, covering approximately 500 hours of continuous monitoring.[10]

Dr. Harold Zahl, then Fort Monmouth's Director of Research, later reported that "we had no legal project set up for tracking Russian satellites ... but within our own laboratory, we had an immediate potential, and duty called desperately." He recalled that a "small select group of Signal Corps R&D personnel at Fort Monmouth," eventually dubbed the "Royal Order of Sputnik Chasers," vowed to work without overtime pay twenty-four hours a day, seven days a week, "until we knew all there was to be learned from the mysterious electronic invader carrying the hammer and sickle."[11] According to National Aeronautics and Space Administration (NASA), the Cold War era launch of that electronic invader "ushered in new political, military, technological, and scientific developments. While the Sputnik launch was a single event, it marked the start of the space age and the US–USSR space race." Fort Monmouth personnel were there when the gun sounded the beginning of that race.

An elaborate (and this time official) monitoring facility was then set up in time to monitor Sputnik II, launched in November 1957. Once again, Fort Monmouth's Deal Test Site was the first American station to record the signals. According to Dr. Zahl, the Command's Technical Information Staff lacked official permission to release the news of Sputnik II's entry over North America. Being 2:50 in the morning, the Pentagon "was not yet ready for the space maze." So, George Moise of the Technical Information Division "quite innocently" called one of the news wire services and asked, "Did anyone hear the new Russian satellite over the US before our reception at 0250 hours?" The very excited gentleman on the other end of the line reportedly exclaimed, "What! No—but, but—you've got a story!" Moise avoided possible reprimand when a coworker, Len Rokaw, managed to awaken enough high-level officials to authorize an official news release to answer the demand for information that had been created.

Deal Test Site Satellite Tracking, circa 1969. (Courtesy US Army Communications-Electronics Command Historical Office, Aberdeen Proving Ground, Maryland)

The United States followed the Sputniks with the launch of its first satellite, Explorer 1, on January 31, 1958. Other satellites quickly followed. Solar cells developed by Fort Monmouth powered the Vanguard I satellite when it launched that same year, on March 17. Vanguard's initial orbit time was 135 minutes. Three minutes after the Vanguard I was launched at Cape Canaveral, Florida, its signals were being picked up at—you guessed it—the Deal Test Site. In its first three years, Vanguard I traveled 409,257,000 miles in 11,786 orbits. It proved itself invaluable in scientific computations. For example, because its orbit was definitely known and charted, it was used by map-makers as a true "fix" in establishing positions of Pacific islands never definitely placed before. It also enabled geophysicists to determine that the earth is slightly pear shaped, not the oblate spheroid previously envisioned. Still more important, it established solar cells as the most efficient and reliable source of electrical power for satellites. On February 21, 1965, NASA reported that the tiny satellite's radio voice had weakened to the point where engineers believed it would never be heard from again after a transmission record of almost seven years. Vanguard I is still in the sky today, however (though not transmitting).[12]

When the first communication satellite, Project SCORE (Signal Communications via Orbiting Relay Experiment) launched from Cape Canaveral on December 18, 1958 and broadcast a Christmas greeting from President Eisenhower to the world, it proved that voice and code signals could be relayed over vast distances—using satellite communication technology developed in part at Fort Monmouth. This experiment effectively demonstrated the practical feasibility of world-wide communications in delayed and real time mode by means of relatively simple active satellite relays, and provided valuable information for the design of later communications satellites. SCORE was an Advanced Research Project Agency (ARPA) project carried out by the Signal Corps with the Air Force providing the Atlas launching vehicle. Project SCORE ceased transmissions December 31, 1958, concluding twelve days of operations and ninety-seven successful contacts. The SCORE satellite was in space for thirty-three days. On January 21, 1959, the satellite reentered the Earth's atmosphere and burned up.[13]

As Project SCORE mechanical engineer John Cittadino recalled,

> you go back in history and Sputnik had been launched, I think it was November of 1957, by the Russians.[14] And there was a mad rush by the United States government to show that the United States could do whatever the Russians could do. And a number of programs were put into effect around the country to get a satellite in space. The first one that actually got up was put together by the people at Redstone Arsenal.[15] Which is now, you know, Huntsville. But it was nothing more than a little ball the size of a grapefruit, and all it did was produce a telemetry signal. You know, it just went "beep beep beep." The job that we were given was to develop an actual communications satellite that would transmit both voice and data. And the laboratory was given the job of designing and building that package. Okay? And we were all told at the start of the game that the purpose was that it was going to be put in orbit. It was to become an active communications satellite. During the time, from the time we started there were several attempts to launch other satellites, notably by the Navy ... they announced that they were going to launch a satellite, and three times in a row they made public announcements, they had the news media down at Cape

Canaveral waiting for things to happen, and the missiles blew up on the pad. It was the Vanguard program.[16] Okay. Well, after this happened three times which caused big embarrassment, as I understand it, President Eisenhower himself said that "we are no longer going to announce ahead of time that we're putting things in orbit. We're going to wait until we have a success." And all of the satellite programs went from unclassified to classified programs. The word came down to Fort Monmouth that our program was cancelled, and that everybody on the project was told that the program was terminated. However, they wanted it to continue, but they didn't want anybody to realize what we were doing. So eventually they brought a small group of people onboard at Fort Monmouth—and you got the number that there was only 88 people in the whole country that were aware of it, and most of them were involved in the building of the missile, which actually carried the satellite. But a small group of us were told that "we're going to continue the program, however, we don't want the word to become public that we're continuing it as a satellite program." So we had to come up with a cover story for the many people at Fort Monmouth. There were probably at least a hundred people involved in working on it, because in that era, we had our own internal shops where we could build almost anything. So there were people in the machine shops and the sheet metal shops that were working on it. So the word was passed that we were going to continue the program ...[17]

The work being done was so audacious at the time, even the team was not sure it would work. When John Cittadino was asked, "So, at the time, were you convinced that it was going to be successful?" He laughed and said, "No!"[18] He reflected that Project SCORE's "legacy is enormous when you consider how commonplace and important satellite communications has become to the world. Not only for the military, but even more so for the commercial world ... So, it was a major step forward." Cittadino did humbly include, "Now, if we hadn't done it, somebody else would have, you know, done it soon after. We just happened to be the ones that did the first one."[19]

Vanguard II, the first experimental weather satellite, launched on February 17, 1959. It also carried an electronics package developed at Fort Monmouth. This Vanguard II, with infrared scanning devices to provide crude mapping of the earth's cloud cover and a tape recorder to store the information, operated perfectly during the entire 20 day life of the battery power source. It made 211 orbits and was successfully interrogated 155 times to release the stored information.[20]

Model of the Vanguard II "cloud cover" satellite. (Courtesy US Army Communications-Electronics Command Historical Office, Aberdeen Proving Ground, Maryland)

The next big satellite milestone came when the labs supervised the development and building of the first high capacity communication satellite, Courier. Courier was a 51-inch sphere, the outer surface of which was studded with 19,200 solar cells. When the sun shone upon them, these generated 62 watts, which could be used immediately or stored up in batteries. The most important items of Courier's equipment were five tape recorders, one for handling voice, and four for storing the ultra-high speed messages. All equipment, including the recorders, four receivers, and four transmitters, weighed 300 pounds. The satellite weighed 500 pounds.

Courier was launched on October 4, 1960. It went into orbit and began to receive, store, and transmit to earth a stream of voice and telegraph radio messages at the rate of slightly more than 67,000 words a minute. Nine days after the launch, a photograph transmitted from Fort Monmouth was retransmitted from Courier and received with no substantial loss in quality. This achievement established the feasibility of satellite storage and relay of all types of facsimile messages, including letters, maps, charts, and photographs. Courier's communication system broke down after seventeen days operation.

Developed and launched as an experiment to test the feasibility of a global military communication network using delayed-repeater satellites, Courier was replaced with a group of three satellites that followed twenty-four-hour equatorial orbits at an elevation of 22,300 miles above the earth. In twenty-four-hour orbits, the new satellites were to follow exactly the rotation of the earth. In effect, each hovered permanently over a particular point of the equator, and the three together were intended to supply an ever-available means of contact between points lying, roughly, between the Arctic and the Antarctic circles. These three communication relay stations were to form the Synchronous Satellite system. To explore and develop this new means of communications, the Initial Defense Communications Satellite Program, known as the IDCSP, was established. The responsibility for carrying out the Army's portion of this program was assigned to the Army SATCOM Agency at Fort Monmouth.[21]

Fort Monmouth further proved instrumental in TIROS (Television InfraRed Observation Satellite) I, the first full scale weather satellite, also developed under the technical supervision of the Fort Monmouth laboratories and launched that same year. TIROS I sent televised weather photographs of the earth's cloud cover and weather patterns to our giant 60-foot "Space Sentry" antenna in the Evans Area. The *New York Times* reported that TIROS also had "obvious military implications" as the orbit of TIROS I permitted it to photograph southern Russia and Communist China. Reports show that the US freely distributed TIROS I's images to those countries both as a goodwill gesture and to demonstrate the advanced pace of American research.[22]

The Fort Monmouth space capabilities at Deal dropped off gradually as NASA and the Air Force set up their own monitoring and tracking facilities. In compliance with Army and DoD directives to abandon excess leased real estate, Fort Monmouth terminated the lease with the Deal site owners effective June 30, 1973. Most Deal facilities and personnel moved to the Camp Evans site in Wall Township (which, as discussed in earlier chapters, the Army had acquired during World War II and still retained).[23]

The contributions of the Signal Corps to the advancement of satellite communications are many. According to former Signal Corps Laboratory Research and Development Laboratory employee John Cittadino in a January 2009 interview:

> its legacy is enormous when you consider how commonplace and important satellite communications has become to the world. Not only for the military, but even more so for the commercial world. I mean, things like broadcasting the Olympics, they do all of that via satellite communications. So, it was a major step forward. Now, if we hadn't done it, somebody else would have, you know, done it soon after. We just happened to be the ones that did the first one.[24]

And satellites were not just a gimmick. As a 1961 news article noted:

> The Signal laboratory is not preoccupied with satellites. While among the more dramatic developments of the era, they are nevertheless not ends within themselves. As shown by their use in communication and meteorology, they are practical vehicles for specific purposes. Many of the advanced electronic parts and systems which go into them have a common use in mundane and not so mundane developments, such as a new family of frontline radio sets the laboratory has developed for soldiers on the ground.[25]

Today, we count on satellites daily for weather information, communications, navigation, and so, so much more. As US Representative from

New Jersey Rush Holt noted in Congress upon the 50th anniversary of TIROS in 2010:

> It's too easy to take for granted the US victory in the space race and the technological developments that were pioneered by TIROS and its successors ... From solar cells and tape recorders to cell phone cameras and GPS systems, the contributions that derive from the TIROS program are not confined to outer space. TIROS is a reminder of what we can achieve when we apply sufficient energy and resources to research and development in pursuit of a national goal.[26]

Fort Monmouth's technological contributions are truly embedded in all aspects of our daily life.

## Other Top Tech

Scientists at the Fort busied themselves with more than satellites in the early Cold War era. For example, the labs played a huge role in the development of the first large-scale mobile computer, the Mobile Digital Computer (MOBIDIC), during the late 1950s and early 1960s. Once referred to as a "portable electronic brain," this van-mounted computer represented the first experiment in automating combat support functions in artillery, surveillance, logistics and battlefield administration.[27] It was "developed as part of a family of systems to store and reduce to meaningful terms the mountains of information essential to today's dispersed, fast-moving army."[28] Fort Monmouth would continue pioneering early military computer technology, and teaching soldiers how to use it. This training impacted not just US Army soldiers, but military personnel from around the globe. As Charles Smith, who first attended and then worked at the Signal school in the early 1960s prior to setting up communications in Vietnam, recalled, "The Army had one of the first UNIVAC computers, big old thing, room-sized, and another room for just the air conditioning ... we had officers come in from all over the world ..." for training on it, to include from places like Thailand, Laos, and Cambodia.[29]

Another focus remained the weather. Fort Monmouth Signal Corps scientists, in conjunction with New York University's College of Engineering, also developed a method to measure rain in 1961. The device, placed outdoors, would size and collect information on all the raindrops falling in a 1¼ inch square. Interesting, yes, but was this necessary? Yes, for numerous reasons, the most important of which being that safety engineers and aircraft designers needed to know the size and number of raindrops to determine the effects of rain on jet engines and the surfaces of supersonic aircraft and missiles.

Radar was still constantly being innovated as well. In 1962, the Fort's experimental ten-pound radar unit used the latest micro-miniaturization technology to spot moving targets more than a mile away. This prototype model took the first step towards the production of light, handheld tactical electronics equipment. This was the first such product that could be held during its operation.[30]

Fort Monmouth even developed a Morse Code Readout device, unveiled in 1965, which plugged into any Army radio and transformed the dots and dashes of Morse Code into letters formed by a light-emitting diode (LED). This device allowed a Soldier with no knowledge of Morse Code to receive coded messages. The ability of the US to decipher German and Japanese codes had played a decisive role in the outcome of World War II, and the importance of cryptography persisted during the Cold War era.

That same year, a single pencil-size laser beam simultaneously relayed all seven of the television channels broadcast from New York's Empire State Building between two points at Fort Monmouth. The volume of information, believed to have been a record, further demonstrated the high capability of laser as a means of relieving over-crowded radio channels. Efficient communications networks were a key to winning the Cold War era race to perfect the best electronics.

These are just a few specific examples of the innovations coming out of the laboratories headquartered at Fort Monmouth. They developed and tested all phases of work to fit its equipment into the new concept of rapidity and flexibility of communications, jamming of the enemy's electronic equipment; utilization of light, mobile rapid computing machines to assist battlefield commanders in making decisions; the use of photographic drones over the enemy's lines, and the employment of silent radar sentries, television, infrared detection and photographic devices and seismic and acoustic systems for battlefield surveillance.

## Pigeons Fly the Coop

Advancements in communications and electronics systems had come so far that in 1957 the Army discontinued the pigeon service, which, as discussed in earlier chapters, had been a fixture on post since the end of World War I. Fort Monmouth pigeons had been used in theater as recently as Korea, where they proved particularly useful to covert operatives in enemy-controlled territory. Their usefulness in this conflict, however, was short-lived. Field Manual 100–11, "Signal Communications Doctrine" (July 22, 1948) stated: "The widespread use of radio in conjunction with the airplane to contact and supply isolated parties has rendered the use of pigeon communication nearly obsolete."

From pigeons to satellites… "Potpie" gets his discharge from Colonel Clifford A. Poutre as the Army prepares to do away with the pigeon service due to advances in communications technology. (Courtesy US Army Communications-Electronics Command Historical Office, Aberdeen Proving Ground, Maryland)

The Department of the Army discontinued its pigeon service in 1957. Fort Monmouth donated its fifteen living "valiant" pigeons to zoos in various parts of the country. The Army sold the remaining birds, approximately 1,000, for five dollars a pair. "Hero" pigeons with distinguished service records were donated to zoos; others sold for five dollars a pair. Newspapers across the country covered the end of the pigeon service, and folks lined up as far as the eye could see, hoping to acquire Fort Monmouth birds of their own.[31]

## Project Paperclip

Some of the pioneering work being conducted in the labs at Fort Monmouth and its sub-installations like Camp Evans during the Cold War era was done by German scientists who had been brought to the US as a part of Project Paperclip after World War II.[32] The idea was to bring these highly skilled personnel to America lest they be commandeered by the Soviet Union. Dr. Harold Zahl, director of research at Fort Monmouth, recalls in his book *Electrons Away* the moment he was asked if he wanted any of this German talent. He said, yes, and reflected, "This was probably one of the most important decisions I have ever made."[33]

Was this ethical? As one 1947 *Asbury Park Press* article explained, on behalf of the military, to locals who might be less than anxious to welcome German scientists into the local community:

> Although these men are under military custody, they are not prisoners of war. They remain under army control until their return to Germany or until they might be placed under immigration visa when they fall into the jurisdiction of the justice department. There are no "big nazis" among them, it was emphasized. It was pointed out that it is contrary to US policy to bring in any one who was an active nazi during the Hitler regime or who would be otherwise objectionable. Among conditions which bar a prospective entrant are: membership in the nazi party prior to 1933, leadership in it at any time, conviction by a denazification board, charges or conviction of a war crime or a criminal record. All are investigated thoroughly before leaving Europe and should any latent Nazism be uncovered after arrival here, the offender is shipped back.[34]

According to one source, twenty-four German scientists were initially assigned to the Signal Corps laboratories at Fort Monmouth in 1947.

By 1955, that number rose to 52.³⁵ One, Hans K. Ziegler, rose through the ranks to become the top scientist at Fort Monmouth and then, in 1963, the top scientist for the entire Army Signal Corps. That same year, the Army awarded him a decoration for Meritorious Civilian Service for his work, which the Army said, "added greatly to the Nation's security and scientific advancement." Ziegler is best noted for his application of solar power to communications and weather satellites.³⁶

In a 1984 article with the *Asbury Park Press*, Ziegler acknowledged that he began paying two dollars a month dues to the Nazi party in 1937 as a condition of keeping his job as an electrical engineer. As he told the paper, "Opposing the Hitler regime would have been suicidal. You either lost your job or were thrown into a concentration camp." He spent the war years near the Czechoslovakian border, developing an artillery proximity fuse that worked on an acoustical principle. A transmitter sent out sound waves, and when the waves returned from a nearby target they exploded the shell. The concept worked, but Ziegler said his fuse never got into production before the war ended. He was recruited by the Americans, who thoroughly investigated his background prior to bringing him to the United States. Both he, his wife and their two German-born children became American citizens in 1954. As he recalled, "America was very good to us. We got unlimited opportunities."³⁷

Ziegler would even vouch for other German scientists seeking to immigrate to the United States in search of such opportunities. For example, in a June 30, 1948 memorandum, he discussed the suitability of one Dr. Ing. Ludwig Grassl, writing, "… it is my understanding that Dr. Grassl is doing outstanding work and is highlight acknowledged in his field. This person's qualifications would certainly make him fit into the picture of SCEL [the Signal Corps Engineering Laboratories]. I have no request from him to favor his transfer here, however I believe he has a certain interest in doing research work in the USA."³⁸

Harold Zahl remained happy with his decision to recruit Germans, writing "It was a wonderful experience to see the old 'Melting Pot' in action … in retrospect, throughout the country we see thousands of our best citizens, able engineers, scientists, and administrators; with a byproduct of tens of thousands of brilliant children in our schools … the results of 'Paperclip.' Surely this country is better and stronger because of that decision …"³⁹

## A Red Scare

If Fort Monmouth's leadership was willing to let German scientists in, it was determined to keep Communists out. Julius Rosenberg, executed with his wife for spying in June 1953, had worked briefly for the Signal Corps Labs during World War II. He was dismissed early in 1945 when it was learned that he had formerly been a member of the Communist Party, but not before he reportedly gave the Soviet Union the secret of the proximity fuse. There was some evidence of other leaks, as well. Fort Monmouth's Commanding General, Kirke B. Lawton, quietly invited anti-Communist crusader Senator Joseph McCarthy (the Chairman of the Senate Committee on Government Operations) to launch an inquiry to prove that Rosenberg, or someone, had created a spy ring that still existed in the Signal Corps labs. Historian Rebecca Robbins Raines notes that Lawton would have preferred to keep his support of McCarthy quiet, however he was ordered to publicly cooperate by the Secretary of the Army, Robert T. Stevens.[40]

Having spent years in very public pursuit of Communists in America, McCarthy was all too happy to investigate the situation in the Signal Corps. He began investigating, and the flimsy charges levied against employees included things like having attended a benefit rally for Russian children, belonging to a union thought to be subversive, and "living with your father," when the father might have Communist sympathies! Some of the employees never even knew what they were charged with. Harvey Lasky, son of suspended employee Solomon Lasky, recalled in an oral history interview:

> Charles Frankel, who was a fine, fine attorney, and one of the more successful practitioners in Asbury Park ... he took many of these cases pro bono. Including my father's case. And he fought the case for my dad ... I remember the *New York Times* calling, I remember, I think, NBC calling, I remember my dad meeting with Charlie Frankel, and Charlie Frankel saying, "don't worry about the money, this is something that has to be done, this is the right thing to do, we will fight this." And I remember that the emphasis, even after—you know, I remember that they kept demanding charges after the suspension. They kept—Charlie Frankel kept demanding, "let me see the charges; what are you charging my client with?" And the charges never came. Then, we—to this day, we never saw any charges against my father. There never were any charges, it was typical McCarthyism— there were just innuendos. He was never charged with anything and I don't think

many of the people—there were some, I suppose, at Fort Monmouth, that were, but not very many. No charges ever came, probably after a year or so my father was reinstated, and then the fight became for back pay, which Charlie Frankel continued to fight for and my father was eventually awarded. He was reinstated and ... he finally did get his security clearance back as well.[41]

The Secretary of the Army began to fear that Lawton was perhaps too enthusiastic in his support of McCarthy's probe as the number of targeted employees grew. Still, Stevens accompanied McCarthy on a fact-finding tour of the post on October 20, 1953. In the hearings that followed, Rosenberg's brother-in-law, David Greenglass, testified from his prison cell that there had indeed been a "Rosenberg spy ring" at Fort Monmouth and as far as he knew, it still existed. But outside of that testimony, McCarthy couldn't find any proof of wrongdoing by any post personnel.

Ultimately, forty-two employees, primarily Jewish engineers, would be suspended for supposedly posing a security risk. Most employees, like Lasky, fought back, with a lawyer for another suspended employee accusing McCarthy of perpetuating "perhaps the greatest hoax in American history"[42] in his probe of the Signal Corps at Fort Monmouth. McCarthy ultimately failed to prove the existence of a Communist conspiracy at the Fort. His actions brought nothing except notoriety to the Signal Corps labs and incalculable grief to the employees who were dismissed from their jobs on mere suspicion. Forty of the employees were reinstated, two instead resigned.[43]

The *Asbury Park Press* encouraged the Secretary of the Army to relieve Lawton of his Command, observing that he "has not always shown the mature judgment expected in a high military post." Lawton would retire that August.

## Country Club of the Army

Despite the brief but very dark cloud of McCarthyism, Fort Monmouth continued to inspire the love and devotion of those who served there. One veteran of this era, William Ryan, recalled:

> I always heard it referred to as the country club of the Army, because it was just that. We, as instructors [at the Signal School], had a good deal more leisure time

than the students did. And certainly more time to do things than troops over at Fort Dix, for instance. I was very active in the community ... one of the young ladies I knew very well ... suggested that it would be nice if I'd come down and join their choir in church in Long Branch, Our Lady Star of the Sea. So I went, I joined the choir, and the next thing I know they're going to put on a stage play. And I worked on the stage crew, and had a wonderful time. And the following year I had a different lady, and she had a similar idea, and I joined the choir in St. James in Red Bank. [Laughs][44]

The post's reputation as a "playground of the Army" was well-known. When Milton Friedrich Langer was in the ROTC at the University of Illinois, he recalled, "You had a choice of Air Corps or Air Force at that time and you had a choice of various branches, the engineers, the infantry, Signal Corps, there might have been some others. If you go back early enough, you had cavalry too. Well, I had a friend that I was in high school with ... and he was in the Signal Corps ROTC and I was in the Air Force ROTC. He told me—Henry told me ... if you go to Fort Monmouth, New Jersey, you can do all kinds of things. There is lots of young women there too, all kinds of things. So he talked me into joining the Signal Corps instead of staying in the Air Force ROTC."[45]

Leo Geisler, who was also stationed at the Signal School during this era, recalled, "Fort Monmouth happened to be 40 miles from where I lived in Plainfield, NJ. So it was an area that I already knew as a youth. And so—it was right near the beach; we had a USO that we could take buses to on the weekends when the weather was nice. And there were dances several nights a week at Fort Monmouth, even when we were going to school, they would import ladies ... it was very nice." Overall, he said, "We had a lot of fun, I was with a nice group of guys ... we just had a good time. I, fortunately, was in a nice area and so ... we did all right. I went in, I followed my orders, and my orders just happened to luck out that way."[46]

It was not all fun and games, of course. This chapter started by exploring how Fort Monmouth played a crucial role in the Cold War race for the best electronics. New generations of radio and radars helped United Nations troops understand the battlefield in the Korean War. Satellite technologies meant that we could increasingly communicate more quickly through time and space than ever before thought possible, and understand phenomena as powerful as the weather. Interestingly, a good deal of the highly scientific work being done at Fort Monmouth was done by people of color, at a time when many workplaces shunned any diversity amongst their employees (outside of menial roles). What made Fort Monmouth different, and how did it come to be known as the "Black Brain Center of the US?" This will be the focus in Chapter 6.

CHAPTER 6

# "The Black Brain Center of the United States"

## Dr. Walter McAfee and His Colleagues Break Barriers at Fort Monmouth

The McCarthy episode discussed in Chapter 5 was a dark period in the Fort's history with regards to civil liberties, to be sure. But in the late 1940s and 1950s, Fort Monmouth also earned a reputation as, in the words of African American electrical engineer Thomas E. Daniels, "the Black Brain Center of the US." Daniels and others affirmed that this post "provided a place where black scientists and engineers could find jobs and advance their careers," while other research facilities closed their doors to African Americans. These civilian employment opportunities did not inoculate the Fort's African American employees against the culture of discrimination and segregation that marked this period in our country's history, however. One should be careful not to paint too rosy a picture. The Army itself institutionalized discrimination until President Harry S. Truman signed Executive Order 9981 on July 26, 1948, ending segregation in the United States Armed Forces. Jim Crow ruled in the private sector, and even New Jersey Ku Klux Klan chapters marched openly in the streets into 1920s. Despite its best efforts, the Signal Corps could not ensure consistent or uniformly equitable treatment for African American Army personnel. Examining the experiences of just a few of the Fort's early African American employees illustrates the dichotomy between the Signal Corps' relatively progressive hiring policies and the day-to-day lives of the African Americans benefiting from them in central New Jersey.[1]

Fort Monmouth and its sub-installations like Camp Evans earned a reputation as the "Black Brain Center" of the United States, 1943. (Courtesy US Army Communications-Electronics Command Historical Office, Aberdeen Proving Ground, Maryland)

Perhaps the most well-known of these talented personnel of color is Dr. Walter McAfee. Dr. McAfee surmounted the racism endemic to twentieth-century America and made unique and enduring contributions to the scientific community during his forty-two years working as a government scientist at Fort Monmouth, to include critical contributions to Project Diana, which allowed man's first "contact" with the moon in 1946. He also made time to quietly battle injustice, and teach and mentor a new generation of innovators and leaders as a professor at Monmouth College (now University) in West Long Branch, a trustee at Brookdale Community College in Lincroft, and an organizer of enrichment programs for high school students.

Walter S. McAfee was born on September 2, 1914, in Texas, one of nine children. His keen mind impressed his high school instructors, one of whom, Freeman Prince Hodge, called him an "intellectual giant."[2] He earned a Bachelor of Science degree in mathematics from Wiley College[3] in 1934, graduating magna cum laude. Still, it was the Great Depression, and he struggled to find work. He did some substitute teaching but was not above odd jobs: farm laborer, carpenter's assistant, door-to-door salesman. Finally frustrated by a lack of options, McAfee recalled in a 1994 oral history interview with Professor Robert Johnson Jr. that "it seemed that I wasn't going to get a job teaching, so I said to my mother and father, 'I'm going to go to Columbus, Ohio, because I can apply for scholarships and fellowships as a first-honor student at the state college.'" He recalled: "Ohio State wrote back and said, 'You will not have to apply separately for admission. You are admitted on the basis of this application.'" Still, venturing to Ohio was a gamble. McAfee set off, "knowing nobody and not knowing where to stay when I first got there. I had $9.29 in my pockets. I asked a black fellow who worked at the bus station if he knew anybody who kept students. He said he knew a very good family and when I got there, I laid it on the line. I said, 'I don't have money. I'm willing to work. I'm a graduate student at Ohio State.'"[4]

McAfee would persevere and earn a master's degree in physics from Ohio State University in June of 1937. It wasn't an easy journey. As Walter's daughter Marsha McAfee Bera-Morris recalled:

> Black people were not allowed to be teaching assistants at Ohio State at that time, nor could they stay in the dormitories on campus. At times, my father was working three or four jobs simultaneously to cover the cost of tuition and housing and educational expenses. His department chairman actively discouraged pursuit of a Ph.D. and, according to my father, hollered at him for not taking an Industrial Arts course. He was, after all, only going to be able to go back South to teach in a black school. [But] Walter McAfee persevered.[5]

In an interesting aside, according to oral tradition, McAfee roomed with and tutored runner Jesse Owens during these years.[6] Owens, you may recall, famously infuriated white supremacist leader of Nazi Germany, Adolf Hitler, by winning four gold medals and setting several records at the Berlin Olympics in 1936. McAfee and Owens's time at Ohio State overlapped. The Ohio State University had only one men's dorm, and Black men were barred from it. Owens's grades *did* at times render him ineligible to compete in collegiate sports. According to student directories held in the archives at Ohio State, Walter McAfee lived at 236 E. 11th Avenue, phone number WA-1570, 1935–1936 and 1936–1937. Jesse Owens lived at

236 E. 11th Avenue, phone number WA-1570 1933–1934 and 1934–1935. The 1935–1936 directory, however, lists Owens at 256 E. 11th Avenue, with the same phone number, WA-1570. The Ohio State archivist and this author feel it highly possible that the 1935–1936 entry for Owens is a typo, and that there is a strong possibility that Jesse Owens and Walter McAfee did at the least share a boarding house.[7]

After graduation, McAfee taught in Columbus, Ohio, from the fall of 1937 to the spring of 1942. During this time, he applied for civil service positions. The widespread racism of the day complicated this process despite McAfee's stellar credentials. As McAfee recalled it:

> Now I was on the list and the first offer I got was Langley Air Force Base in Virginia to solve problems in hydroaeromechanics. According to the paperwork, I was to supply information and send the application back. So I read the duty and at the end it asked me questions about race and religion and to send a picture and so forth. I knew that the NAACP and others had been fighting to get that kind of request removed from applications … We had a friend who was a black chemist and he was suing all the time because he was one of these guys that was always number 1, 2, or 3 at the top of the exams. Whenever he went for an interview, he ended up dead when he sent his picture. At that time, they didn't have to take the top man; they had to take one of the top three.[8]

Knowing the deck was stacked against him, McAfee applied for the Langley job. He recalled: "So I sent it with the picture. They sent it back, 'no'—they had hired somebody for that position. The next offer I had was to teach elementary calculus to the aviation cadets at Kelly Field in Texas. I didn't get that one. It asked for pictures and so forth …"[9] This process repeated itself several times, until a job came up with the Army Signal Corps at Fort Monmouth. McAfee recalled that Fort Monmouth's application did *not* ask for a picture. He received a job offer shortly after applying, with instructions to report almost immediately after submitting his paperwork. He resigned from his steady teaching job in order to do so, despite fears that he might be fired when he arrived in New Jersey and fort officials discovered his race. This was an especially large gamble, as McAfee had married his wife, Viola, at this point, and she was expecting their first child.[10]

McAfee's fears dissipated when he arrived at Fort Monmouth and found a number of African American civilians already at work. The Signal

Corps at Fort Monmouth had in fact been offering unique opportunities for people of color throughout the 1940s and 1950s. African American electrical engineer and retired senior executive staffer Thomas E. Daniels, who worked at the post for thirty-five years, summed this up in a 2003 interview with journalist Gloria Stravelli when he stated: "Fort Monmouth was known as the Black Brain Center of the US" Daniels affirmed that the post "provided a place where black scientists and engineers could find jobs and advance their careers," while other research facilities closed their doors to African Americans.[11]

It's interesting to note that at least some active duty military personnel remembered the post during this period in a similar fashion. Albert C. Johnson, who would become the first Black colonel in the Army Signal Corps, recalled that when he and 21 other young black men arrived at the Fort in January 1943, the post commander, Major General George Van Deusen, told every company commander that there was to be no segregation. The men were to be "scattered around," as Johnson recalled it, noting, "That worked out very well and I began to admire Fort Monmouth very much ... this was one of the few posts that I was assigned to where everybody was considered as good as everybody else, and everybody worked in harmony trying to get the best things accomplished." In another instance, when going to see a movie off-post with a white friend, Johnson was told he had to sit in the balcony. Neither Johnson nor his friend took kindly to this. They conducted a little experiment, trying to sit together at movie theaters in other local towns. They encountered the same problem. They promptly reported this to General Van Deusen, who reportedly called all the local mayors and told them that they would either treat his African-American personnel as equals, or face a boycott by all the thousands of personnel under Van Deusen's command. Johnson remembered Van Deusen fondly, saying "I became very proud of him." As Johnson's career progressed with service in World War II, Korea, and Vietnam, he was always happy to return to Fort Monmouth.[12]

Of course, the civilian employment opportunities noted by McAfee and Daniels and the fair treatment Johnson enjoyed were not a magic wand erasing the prejudice and racism so endemic to the era. Let's return to our primary case study, Dr. McAfee. Upon arriving in New Jersey in spring 1942, McAfee noted that off-post segregation and discrimination made it difficult to get housing and meals. In 1948, another Black Signal Corps employee, Leroy Hutson, had a cross burned in front of his home in Wall Township.[13] Racism and discrimination could rear its head on post as well, sometimes making it difficult for African Americans to receive promotions. As McAfee later described it: "... if they had one position and had a black man and a white man competing for it, the white man got it. Mainly, there's less friction that way. The black man isn't going to fight that hard. Of course, today you wouldn't say that. I guess we were just getting into the jobs."[14] This is not to suggest McAfee was a pushover, by any means. He once said, "I wouldn't work for Jesus Christ if he wouldn't listen to me."[15]

McAfee's early, World War II-era assignments in the Signal Corps laboratories included duty with the radar siting group and the mathematical analysis group. Projects were often top secret and included research into locating mines, counter-mortar equipment, and identification friend-or-foe technology—or, as McAfee described it in lay terms: "Let's find out who they are before we shoot them."[16] Despite the urgency of wartime operations, McAfee recognized that his work could help save lives, and therefore the need for quality control was great, noting: "I always say it doesn't do much good to rush work. You find errors that you didn't think you could make."[17] The project for which Dr. McAfee is perhaps most famous occurred just after World War II—the Project Diana moon bounce, a 1946 radar experiment at a Fort Monmouth outpost then known as Camp Evans in nearby Wall Township, New Jersey.[18] Military brass wanted to determine whether the ionosphere could be penetrated by radar, in order to detect and track enemy ballistic missiles. As the National WWII Museum ably tells it: "The Pentagon ordered ... Camp Evans staff to investigate if such a weapon was launched against the United States whether it would be possible to detect and track it using radar ... the team-modified radar equipment already on hand at Camp Evans for their equipment, using a heavily modified SCR-271 radar set as their transmitter."[19]

Because there were no incoming missiles to track, the team decided that they would try to bounce a radar signal off the moon. Early calculations on how exactly to do this were not working, until McAfee was brought in to puzzle them out. As he recalled it:

Dr. Walter McAfee in his laboratory, 1946. (Courtesy US Army Communications-Electronics Command Historical Office, Aberdeen Proving Ground, Maryland)

> When they came to speak to me initially, they knew that I had done radar coverage diagrams. I had done radar sighting. I had done radar echoing areas or radar cross sections and I had done refraction studies in the atmosphere. I had a paper on it ... Colonel John Dewitt was head of Evans and he had previously tried bouncing radar signals off the moon and failed. E. K. Stodola was head of the civilian branch section that we in theoretical studies were under at that time. I computed a radar cross section of the moon, a radar coverage pattern, and distance to the moon, so they could tell how big the signal would be when it returned.[20]

On Thursday, January 10, 1946, at 11:58 a.m., using McAfee's calculations, the team detected the first signals reflected back from the moon. The radio waves took 2.5 seconds to travel to the moon and back. As the WWII Museum notes:

> The experiment was repeated over the coming days and months and demonstrated for Pentagon officials, who were also interested in the "Moonbounce" technique's potential to eavesdrop on the Soviet Union; experiments which ultimately proved unsuccessful ... Despite its limited military potential, Project Diana witnessed the birth of radar astronomy, or the ability to observe and measure the distance of nearby astronomical objects by analyzing their reflections. Project Diana also demonstrated that radio communication could be conducted through the ionosphere, paving the way for the development of satellites and ultimately manned space-flight. Perhaps more familiar, the "Moonbounce" technique—known today as EME or Earth-Moon-Earth communication—is still used by amateur and HAM radio operators to this day.[21]

The January 25, 1946, *Asbury Park Press* newspaper read:

> Engineers and astronomers saw in the first direct contact with the moon, the opening of a vast field of speculation, which did not preclude the possibility that man might some day journey to the moon ... The radar contact also opened up an entirely new field of study of the universe. The success of the engineers was seen as opening the door to direct communications with other planets in our system."[22]

This was not just press hype, or journalists caught up in the moment. Major General Walter E. Lotz Jr., former commander of Fort Monmouth, affirmed in 1971: "We can note that it was the role played here that made the Apollo 14 and other moon shots possible because it established the feasibility of radio communications throughout outer space." He continued: "Without space communications, our moon probes would never have gotten off the ground."²³ McAfee himself appreciated the significance of Project Diana even at the end of his long career, telling Professor Johnson in 1994:

> A lot of people say that the work I was very much a part of, bouncing and receiving radar signals off the moon, heralded the US's ability to send manned spacecraft. I don't discourage those comments ... If people say that this heralded the space age, then I say fine. You take one step at a time and since we had learned to do radar-coverage patterns and echoing areas and signal to noise ratios and ranges on things, the next thing to do was to make one that had the right parameters to reach the moon."²⁴

While the team leaders involved in Project Diana were happy to boast openly about their momentous accomplishments to the press and general public (the existence of the project was not classified), credit was not shared properly, at least at first. As McAfee recalled: "The press release was sent out but I was not mentioned ... There were a lot of [other] technicians who cried that they weren't named in that paper and should have been, too."²⁵ Several months went by before this oversight was corrected and McAfee was given credit for his work. He recalled:

> Somebody connected with publicity at the army was talking with me and said, "I looked at all those early things and I don't find you anywhere. Can you tell me why that is?" I said, "I can tell you why, but I don't care to get into a discussion about it." I didn't get any of the publicity until after I got the Rosenwald Fellowship to Cornell, which was announced around May or June of that year [1946]. The moon radar was announced in January and then in May or June they wrote big articles about me and then they made some nice statements about how I had worked on the theoretical problems connected with radar. I said, "They could have told that in the beginning."²⁶

Despite this negative experience, McAfee stuck with the army and earned a doctoral degree in nuclear physics from Cornell University in 1949.²⁷ He continued working at Fort Monmouth, on projects including those related to nuclear technology and satellites. In 1956 he was awarded one of the first Secretary of the Army Research and Study Fellowships, which was presented to him at the White House by President Dwight D. Eisenhower himself. Under the fellowship, he studied radio astronomy and ionospheric theory at Harvard University.²⁸

Diana radar, undated. (Courtesy US Army Communications-Electronics Command Historical Office, Aberdeen Proving Ground, Maryland)

It was in the 1950s that US Senator Joseph McCarthy brought his communist witch hunt to Fort Monmouth, convinced a ring of spies was operating at the post. Dozens of employees were suspended, their reputations battered, on little to no evidence.[29] McAfee himself was not the subject of any investigations (that he knew of), but he recalled how he tried to help one man in the crosshairs:

> Only once did I get involved early to help someone. I heard a guy tell that when he was a little boy, he grew up as a Catholic and he was an altar boy. I remembered it, and later when he said they were after him, I said they are fighting any way they can; you fight back that same way. Write back to them and tell them that you grew up as a Catholic and that you were an altar boy when you were a young man and that you subscribe to the Catholic Church's position on communism. He looked at me for a moment and I said, "Write it and send it in." He said, "I seldom go to church." I said, "They don't give a damn. They don't know." So he went back and wrote it. He didn't get put out of there.[30]

It seems McAfee tackled communist hysteria in the same cool, analytical way he regularly went up against racism.

In the 1960s, McAfee developed sensors that were used to detect and track enemy movements during the Vietnam War. In 1961, he won an Army Research and Development Achievement Award "for studies vital to the national defense in connection with missile guidance systems and communications links."[31] In 1969, when America first put a man on the moon, the press wanted to hear McAfee's thoughts. The *Asbury Park Press* reported:

> Dr. Walter S. McAfee, who was among Ft. Monmouth scientists at Evans Signal Laboratories, Wall Township, to make moon contact by radar in 1946, said, "I think it's very wonderful. I think at this stage it's a real achievement and I'm waiting for them (the astronauts) to get back in and get off the moon. I was very impressed with their first statements."[32]

In 1971, McAfee was one of the first African American employees of Army Materiel Command to be promoted to GS-16, a "super-grade" civilian position, and predecessor of today's federal senior executive service. He and Fort Monmouth command leadership traveled to Washington, DC, for the prestigious promotion ceremony.[33] At that time, McAfee became the first scientific advisor to the Deputy for Laboratories at the US Army Electronics Command at Fort Monmouth (as the Signal Corps Laboratories was at that point known). The *Asbury Park Press* newspaper reported:

> The appointment of Dr. Walter S. McAfee of South Belmar as the Army Electronics Command's first scientific adviser to the Deputy for Laboratories will evoke praise from his Fort Monmouth colleagues and equal commendation

from the Shore community where Dr. McAfee has been identified with a wide variety of civic activities for the more than quarter century he has resided here. An outstanding astrophysicist, his educational and professional achievements have earned him a multitude of honors, too lengthy for listing here. A mild and modest man ... it was his theoretical calculations which provided the groundwork for man's first radar contact with the moon in 1946 and his latest adventures are in the rare areas of quantum optics and laser holography. In them he has the admiration of those who do not know what the terms mean but know that the man himself is a valued and concerned citizen of the Shore community.[34]

From 1979 until 1983, Dr. McAfee was an advisor and study director for NATO forces in Europe. "This meant I had to make frequent trips to London, Brussels, Paris, and The Hague," he said in a January 1985 interview with journalist Edward L. Walsh. "All of the officials of the countries involved spoke English, but the French insisted that we had to use their language in all our dealings. As a result, I got to know a little bit of French."[35] This may have been an especially poignant post for McAfee, who, when traveling stateside early in his government career, was often mistaken for a cook or a laborer, and had been required to travel with *The Negro Motorist Green Book*, a "guidebook for African American travelers that provided a list of hotels, boarding houses, taverns, restaurants, service stations and other establishments throughout the country that served African American patrons."[36]

McAfee retired from civilian service with the army in 1985 after forty-two years of service, having spent his entire career based at Fort Monmouth. In an *Asbury Park Press* interview at the time of his retirement, he called his work with Project Diana the "highlight" of his career.[37]

Though he had left his full-time teaching career behind to join the civil service back in 1942, McAfee had remained involved in education throughout his years at the fort. Despite the demands of his flourishing career with the army, he still found time to support and educate the scientific community's next generation of leaders. He was especially interested in helping learners from disadvantaged communities. As early as 1946, just a few years into his army civilian career, he was organizing collections to raise funds for African American college scholarships.[38] He was a founder and an active participant in a Monmouth County group that mentored and tutored local high school students from underprivileged backgrounds. As another mentor, Thomas Baldwin, noted: "The size of public school classes makes it impossible for teachers to give individual attention to every child to the extent that some may need it. Hence, we try to provide that extra attention some students need." Mentor LeRoy Hutson continued:

> Many of these children most of whom are Negroes have never had any intimate contact with Negroes who have come from the same or very similar backgrounds, but who have succeeded by persisting in their school work and have gotten good jobs in professional fields ... The students have few opportunities to come in contact with people who have surmounted their neighborhood surroundings. When we talk to them, we try to motivate them to do so.[39]

McAfee also served as a trustee of the local community college, Brookdale, in Lincroft, starting in 1970 (shortly after Brookdale opened in September 1969). When his candidacy was announced in 1969, the *Red Bank Register* reported: "Dr. McAfee would be the only black person on the board."[40] Upon joining the board in 1970, the *Register* declared his "credentials are impressive—and Brookdale is fortunate to have another outstanding member of its Board of Trustees. Together with an excellent administration and a first-rate faculty, the leadership is being provided to make Brookdale the type of college that is the goal of all of us in Monmouth."[41] McAfee eventually rose to board of trustees chairman, serving in that capacity for five years beginning in 1975. The early years of the college, including McAfee's tenure, were plagued by growing pains, to include battles over faculty pay, academic freedom, and campus growth.[42]

Dr. McAfee also lectured in atomic and nuclear physics and solid state electronics at nearby Monmouth College (now Monmouth University) in West Long Branch, New Jersey, 1958–1975.[43] He was firmly ensconced in student life, regularly joining clubs for lectures and other activities, and serving as a faculty charter member of Sigma Pi Sigma, Monmouth College Chapter (the National Physics Honor Society).[44]

McAfee's classes were the stuff of legend among students, with one course being referred to as the "Mystery Hour." His classes were known to be grueling, but fair. Of the grading, one former student, George Morris, recalled in a recent oral history interview with this author:

> Well, it was curved, of course, it had to be because no mortal that I know would get 75 to 80 or whatever it is. The curve was, if you got around a 40 or 50, you generally got a B. His exams, they probably made up what he thought was reasonable in his universe. He invited you to bring anything into the exam that you wanted to. Any book, any paper, any test, previous tests, anything. He wanted you to succeed. He wasn't there to trick you, although sometimes it looked like you never saw some of the questions before ever in any of the homework.[45]

Another former student, Ron Johnson, concurred about the rigors of McAfee's courses, recalling in his recent oral history interview with this author: "One of the tests I took, I think I got two or two and a half points out of 100. Very depressing, by the way. I looked, and I had a B on my paper for that. And they always had the rumor that he gave you half a point for putting your name on the paper, so I don't know, but it was tough!"[46]

But the students seem to have loved and respected McAfee. Gary Barnett noted in an interview with Monmouth University student Vincent Sauchelli: "I had him for advanced physics. He was a very interesting man ... He was really, really brilliant."[47] John Tranchina, in his interview with this author, praised Dr. McAfee's teaching ability, saying:

> He would write everything on the board and you didn't have to scramble ... and miss things. So if you didn't understand something, you could read it, right? Again, and ask him right there. And he was right on the money all the time. He was very rigorous in what he taught. And I learned a lot from him, I mean, he was just a great teacher. I had some terrible teachers in college, but everybody gets those, and you have to work at writing, scribbling and everything ... You didn't have to do that with him, you just ... It was great ... What he did was taught down at your level.[48]

George Morris noted: "I think in terms of working, I probably put more time into that course ... You know what's interesting? I did it because I wanted to please him. He had a nature about him that wasn't adversarial. It was something that—it became friendly, and that was great." Morris continued: "He didn't look down his nose at you because you weren't as smart as he was. He understood that you're an engineering student ... and [would] help you in any way he could to enjoy what he enjoys and try to convey to you the fun of learning things that you'd never thought you could learn before."[49] Many of the students McAfee helped educate at Monmouth College would head over to nearby Fort Monmouth and have long, productive, and important careers with the army themselves. Many recalled seeking out their former teacher while at the post, remembering that McAfee still readily dispensed advice and support long after they'd moved from teacher and pupil to colleagues.

Dr. Walter McAfee died on February 18, 1995. He was survived by his wife, Viola Winston McAfee, and their two daughters, along with many other beloved family members. His awards and commendations are too many to list comprehensively here.[50] He has been memorialized in some distinctive ways by the communities he contributed so much to. For example, in 1997, Fort Monmouth named a $14 million building in his honor. It would house the Communications-Electronics Command's Information and Intelligence Electronic Warfare Directorate (a descendant of the old Signal Corps Laboratories, and of the Electronics-Command). This was a unique designation, as most memorialization on army posts honors uniformed military personnel. Then Communications-Electronics Command Commander Major General Gerard P. Brohm noted at the dedication ceremony:

> This building will stand through the future years as a physical monument to Walter S. McAfee—to his intelligence, his graciousness, his courage, his accomplishments, and his caring for all those who worked with or for him. But the true monument to Dr. McAfee will always live in the hearts and minds of the people whose lives he touched and who were enriched by their relationships with him, and that legacy will outlast even this building.[51]

US Representative from New Jersey Michael Pappas (R-12th District) observed that the memorialization was appropriate as:

> Dr. McAfee's accomplishments are a testimony to our Nation's unrelenting thirst for knowledge and his spirit lives on in our national space programs. Our country would not be where it is today if it were not for the creative minds and work ethic like that of Dr. McAfee.[52]

McAfee's widow, Viola, told the *Asbury Park Press*: "He would have loved knowing that this building was named after him, it would have made him so happy."[53]

When the Army's Base Realignment and Closure Commission closed Fort Monmouth in 2011, the Communications-Electronics Command named another building for McAfee, at their new campus at Aberdeen Proving Ground, Maryland. Velma McAfee-Williams, McAfee's only surviving sibling, attended the dedication ceremony and toured the McAfee compound. She observed: "It's a great honor that his line of work and his contributions are being recognized here. This was such an enlightening and overwhelming experience. The tour was outstanding, and I got to go inside various labs and see different programs being worked on in this space." Army Public Affairs Officer Kristen Kushiyama reported that McAfee-Williams said while her brother loved Fort Monmouth and probably would have been sad to see it close, he would have appreciated the larger space and facility.[54]

Then, in 2015, McAfee became the first African American to be inducted into the Army Materiel Command's (AMC) Hall of Fame.[55] The AMC Command Hall of Fame, established in 2012, "honors and memorializes those soldiers and civilians who have made significant and enduring contributions to AMC and the Army. The Hall of Fame preserves the command's history and recognizes the exceptional leadership, service and dedication of former AMC members for their remarkable efforts."[56] Historian Susan Thompson, who nominated McAfee for the honor, declared: "Dr. McAfee is one of those individuals whose contributions, both in communications technology and as a mentor to others, makes him an exceptional figure in … history … His personal and professional qualities should be more widely appreciated, and therefore, I thought him an ideal candidate for the AMC Hall of Fame."[57]

And in 2019, the US postal office at 1300 Main Street in Belmar was renamed in his honor.[58] US Rep. Chris Smith (R-4th District), who introduced the legislation to dedicate the post office building to McAfee, observed: "We remember and honor him for his lifelong commitment to learning, including his service as chairman of the board at Brookdale Community College." He also acknowledged McAfee's perseverance in achieving his goals and pursuing his dreams, noting: "As an African American, Dr. McAfee overcame adversity and prejudice with courage, tenacity, and faith … His amazing life inspires. He challenges us to strive for excellence. He is a true role model."[59] McAfee's youngest daughter, Marsha McAfee Bera-Morris, spoke of how fitting it was that her father be commemorated in such a public space in the area where he long resided and raised his family, saying: "This moment provides a singular pride and satisfaction for so many of his family and friends and colleagues."[60]

Recently, Monmouth University established a scholarship fund in Dr. McAfee's name to "support economically disadvantaged students to attend Monmouth University in pursuit of an education in any of the sciences, while celebrating a distinguished faculty member who broke racial and scientific boundaries."[61] Dr. Walter Greason, who has written extensively about New Jersey and particularly the area around Monmouth University, expanded on the significance of having McAfee at Monmouth in a February 2021 scholarship fundraising event, noting that the region around the college "… wasn't always a place dedicated to equal rights and equal justice for everyone … In the 1920s … the Ku Klux Klan dominated the Jersey Shore, so just a generation later for someone like Dr. McAfee to become a major contributor to … Monmouth College … is a major breakthrough."[62] The university pointedly decided to establish a scholarship before naming anything on campus for him, believing that investing in the education of students would be the best way to honor his legacy.

Dr. Walter McAfee and the other pioneering scientific personnel of color at Fort Monmouth overcame the racism endemic to twentieth-century America and made unique and enduring contributions to the scientific community. McAfee dedicated nearly his entire adult life to government service, while still finding time to mentor the next generation of scholars and slowly chip away at societal injustice. While the communities to which he contributed so much have memorialized him in different ways, the general public knows too little about his accomplishments and one hopes you, readers, can help to spread the word.

It's noted above that McAfee developed sensors that were used to detect and track enemy movements during the Vietnam War. In our next chapter, we'll look at how McAfee and his colleagues on post, in innovating such devices, became "The Eyes, Ears, and Voice of the Fighting Man" during that conflict.

CHAPTER 7

# "The Eyes, Ears, and Voice of the Fighting Man"

## The Vietnam War and Its Aftermath[1]

Even as the Korean War ground to a bloody stalemate in 1953, it was becoming clear to those following world affairs that Southeast Asia was a powder keg. As the French abandoned their grip on the region following their stunning loss at Dien Bien Phu in 1954, the United States become increasingly entrenched in supporting the South Vietnamese (in support of the domino theory which posited that if one country fell to communism, all of its neighbors would follow). The number of American advisors on the ground, the first of which arrived in 1950 during the administration of Harry Truman, ticked steadily upward before giving way to massive bombing campaigns and ground troops in what would become a painfully divisive war.

Brigadier General Walter E. Lotz, Jr., Assistant Chief of Staff for Communications-Electronics (ACSC-E), US Army, Vietnam (USARV), stated in 1966 that "Electronics has never been so vital in a war as it is here in Vietnam."[2] During the conflict, organizations headquartered at Fort Monmouth trained men to use electronics at war **and** managed signal research, development, and logistics support, supplying combat troops with a number of high-technology commodities. These included mortar locators, aerial reconnaissance equipment, sensors, air traffic control systems, night vision devices, and surveillance systems.

Max Cleland, who served as an aide to the commandant of the Army Signal Center and School at Fort Monmouth General Thomas Matthew

Rienzi, recalled that the enormity of the responsibility placed upon Fort Monmouth in the Vietnam era was palpable. He recalled, "This was war and we were training young men to go to war. And that was serious. And ultimately, Fort Monmouth went to a 24-hour-a-day seven-day-a-week operation. And you had … young signalman, you know, going to Vietnam, so it was a—it was a very, very aggressive, go-go-go, upbeat, we're in a war, we've got to take care of our troops kind of atmosphere."[3] Of course, despite the world-class training and technological edge, the Americans, like the French, ultimately withdrew from the quagmire in Vietnam. The last American combat troops left in March 1973 and Saigon fell in April 1975.

## Vietnam War Era Changes on Base

Fort Monmouth executed its Vietnam War era missions in the midst of enormous and sometimes seemingly arbitrary changes on post. Since the installations' founding during World War I, the main activities on post had been the Signal Corps school and the Signal Corps laboratories (both operating under a few, slightly different names over the years). The 1960s brought truly significant changes to Fort Monmouth, though. On February 16, 1962, Secretary of Defense Robert S. McNamara's "Project 80" was quietly put into effect. Although done with little fanfare, Project 80 totally reorganized the Army. It abolished the technical services (like the Signal Corps) and assigned their schools to the Continental Army Command. It created a Combat Developments Agency to handle Army doctrine. And it established a single, giant, logistics agency, the Army Materiel Command (AMC), to handle all logistics, research, and development for the Army.[4] This might sound a bit complicated to the casual reader, but basically it totally changed the structure of the Army and had ramifications for how Fort Monmouth (and many other installations) operated.

One of the new major subordinate commands of the newly created AMC, the US Army Electronics Command (ECOM), was activated at Fort Monmouth over the spring/summer of 1962 to handle most of the logistics functions that formerly belonged to the Office of the Chief

Signal Officer, as well as all associated organizations, installations, and personnel.[5] This included, at this point in time, a work force of about 14,000 people (not all of whom physically worked at Fort Monmouth or even in the state of New Jersey) and a budget of $760 million. Stuart S. Hoff was appointed the first Commander of ECOM.[6] Changes to the organizational structure of this new Electronics Command would continue for the next several years. Basically, though, the ECOM would continue in the tradition of its Signal laboratory forbearers and supply combat troops with a number of high-technology commodities during the Vietnam conflict. These included ever smaller radios, mortar locators, night vision devices, and surveillance systems. We'll discuss the labs more in just a bit.

But what of the Signal School, which had long been so central to the identity of Fort Monmouth? Was it a part of ECOM? During the Vietnam War, the majority of the Signal Corps enlisted personnel trained at the Southeastern Signal School at Fort Gordon, Georgia, not at Fort Monmouth. Fort Gordon, not Fort Monmouth, also hosted the Officer Candidate School. The Army would soon order the consolidation of the Signal School activities split between Fort Monmouth and Fort Gordon. This consolidation was meant to economize manpower and operational costs for communications–electronics training. It was thought that combining the two schools at Fort Gordon would provide greater efficiency in the administration and support of academic programs, and a year-round climate more conducive to the conduct of field exercises (think far fewer snow days). The Army also claimed that Fort Gordon had better access to adequate field training sites.[7]

It was true that Fort Monmouth's main post was at this point hemmed in in the midst of a rather heavily developed area. One Cold War era veteran, Frank Effenberger, recalled in an oral history interview that during training:

> part of it was to go out in a night march and I could hear the juke boxes playing, you know, from Route 35 and the bars there ... and we're marching practically through people's back yards, and of course they all want to know what was going on. The last night of the bivouac they held a mock battle out in the woods. We were all issued blank ammunition ... they had a machine gun setup firing and making a big racket. Police pulled into the clearing and wanted to know what was going on ...

One could see how music spilling out from the bars and police interrupting your maneuvers could be a bit distracting!

Not everyone agreed with the decision to move the school, of course. Fort employees and the local communities formed a "Save our Signal School Association." An editorial in the *Asbury Park Evening Press* observed, "Fort Monmouth's school has made notable contributions to the defense effort. Its civilian force has frequently adjusted to recurring expansion and retrenchment dictated by military requirements." Representative James J. Howard declared the move "A waste of the taxpayers' money and an insult to the people of the Third Congressional District." State Senator Joseph Azzolina called the idea, "Typical false economy." Assemblyman Joseph Robertson said, "If that's the Army's idea of economy, we're in bad trouble."[8]

Veteran William Ryan, who trained at Fort Gordon and later taught at the Signal School at Fort Monmouth, and thus had experienced both installations, said of the move, "I thought it was the worst thing they ever did. They moved, essentially, the core of talent away from the center. The ... center of communication industry, is New York City. Or California. But to take it and move it all the way to Georgia, away from that reality, I thought was a disaster."[9]

But the Army clearly felt their calculations outweighed these opinions. On July 1, 1974, the Southeastern Signal School at Fort Gordon became "The US Army Signal School." The Signal School at Fort Monmouth continued to operate for a time as "The US Army Communications-Electronics School," while equipment and personnel transferred. Fort Monmouth's last class in signal communication graduated on June 17, 1976. Private First Class Rose Hull had the distinction of being the last of some 280,000 servicemen and women of all ranks and all arms and services to receive a diploma from the school.[10]

The tried and true pairing of Signal School and laboratory, formalized in 1919, was broken up with the official closing and transfer of the school to Fort Gordon. The movement of the school involved the transfer of only 89 civilians who opted to accompany the school. More than 700 others were either reassigned to other agencies on Fort Monmouth or retired. The Signal School at Fort Monmouth had existed under various

# the monmouth message

published in the interest of Fort Monmouth personnel

June 16, 1976
vol. 29-no. 4

circulation 12,000

*Signaling the end of an era*

## School to hold last graduation

The former Army Signal Center and School — redesignated the Army Communications Electronics School in 1974 — will graduate its last class tomorrow, ending more than 56 years of continuous training in signal communications at this post.

Renowned as "The Home of the Signal Corps," the famed Army school has been consolidated with the Army Signal School at Fort Gordon, Ga. It is scheduled to close its doors October 31.

Even as the school prepares to close, its training policies reflect its approach to up-to-date military, educational and social trends. Its last graduate is a woman.

PFC Rose M Hull, 20, of Elizaville, N.Y., has the distinction of being the last of some 280,000 servicemen and servicewomen of all ranks, and of all arms and services, to receive a diploma from the school. Her Military Occupational Specialty (MOS) is indicative of the advancement in communications technology since the school's first classes in telephony, telegraphy, and early radio began in 1919. She will be graduated from the Automatic Digital Message Switch Repair Course, and has been posted to Coltano, Italy. In addition to PFC Hull, 17 other students will graduate.

Tomorrow's exercises begin at 9 a.m. in Myer Hall auditorium. The public is invited. Col. Earl R Weidner Jr., Army Tactical Data Systems Agency (ARTADS), will be the speaker. Prior to his ARTADS assignment, Weidner was the director of training and education at the school. The 389th Army Band will play.

Although Oct. 1, 1919, marks the formal establishment of the Signal School at this location, training in some courses date back to 1917 when Fort Monmouth was known as Camp Alfred Vail. At that time the curriculum included physical training, dismounted drill, pitching tents, first aid, cryptography, heliograph, semaphore, and wigwag. Tomorrow's ceremony will phase out the last of 75 courses devoted to training personnel to operate, repair, manage, and engineer sophisticated satellite and computerized systems for worldwide communications.

When the Signal School officially was established here in 1919 its 225 students attended classes conducted in four large hangars, built in 1917 to house aircraft and equipment used in airplane direction-finding experiments. One of the hangars is still standing, housing the Field Print Plant on Oceanport Avenue. In the '60s, during the buildup for the Southeast Asia war, the school was utilizing 500 classrooms in 175 separate buildings. Classes were being conducted on a three-shifts-a-day basis, five days a week. In 1970, a peak year, 13,228 enlisted personnel and 1,788 officers passed through the school. A combined military and civilian staff and faculty numbered 2,500.

The affiliation of the Signal Corps with the transmitting radio amateurs of the country in 1926 was cemented with an Army headquarters station located at the Signal School. The greatest distance range of the radio sets used

The last of some 280,000 servicemen and service women of all ranks, and of all arms and services, to be graduated from the former Army Signal Center and School, PFC Rose M. Hull, 20, Elizaville, N.Y., makes final run-through on training equipment. A 1974 graduate of Germantown (N.Y.) High School, she is the daughter of Mr. and Mrs. Dmetre Hull Sr., Elizaville.

in 1926 and taught at the Signal School was 11,820 miles to a point in Australia.

Three or four weeks ago the school graduated its last class in satellite terminal repair. Former graduates of that course have provided backup communications for the Apollo moon mission and have participated in launching the first synchronous communications satellite.

The advent of World War II resulted in more extensive demand for communications specialists. An Officer Candidate Department was activated in 1941 and 21,000 second lieutenants were graduated within the first three years. In addition, the Replacement Center here, which by 1941 had already trained 13,000 enlisted specialists, trained some 45,000 more by war's end. Outbreak of hostilities in Korea in 1950 accounted for another big buildup and upwards of 11,800 men were trained in the first 11 months of 1951.

Over the years the Signal School also has trained several thousand foreign officers and enlisted men representing 66 different countries. In July 1961 an Iranian Army captain had the distinction of being the 50,000th graduate of the then Officers' Department. Last year as the war in Southeast Asia ended, five Vietnamese and one Cambodian were enrolled in the Communications Electronics Systems Engineering Course.

Until recently, the Signal School administered some 260 correspondence courses to a yearly average of 15,000 registered students at home, or on duty stations in every part of the free world.

**World Acclaim**

Recognized by educators as one of the world's leading technical institutions, the Signal School pioneered many innovations in training and education. The effectiveness of its Instructor Training Branch, affectionately known as "the Charm School," was highly regarded in both military circles and among those state and national public officials who had training responsibilities. State police, lay teachers in vocational schools, and Boy Scout leaders were among those who took advantage of its learning techniques.

Among many "firsts" researched and developed at the Signal School was a system of programmed learning. Many of the self-tutor programs used in classrooms around the world were developed here. It was the first Army school to use television for instructional purposes. As early as 1951 it operated a closed-circuit TV system, which, when phased out last October, was the second largest military closed-circuit TV system in the world. During training for Vietnam, the school's TV Division operated as many as 21 channels at once, 24 hours a day.

It reached 500 classroom outlets, and five auditoriums equipped to project images on motion picture screens. It was beamed into Patterson Army Hospital, enabling convalescent soldiers to keep abreast of classwork.

The Signal School met the challenge of the electronics age. Its basic and advanced courses covered the technologies of 215 major types of electronic equipment from radar sets and from the earth-based elements of satellite communication equipment straight up the scale to computers. The courses also covered all of observational meteorology including what was taught in no other school — micrometeorology. And they covered motion-picture photography, still photography, and photographic laboratory operation.

In the early '50s, the training world beat a path to the door of the Signal School to observe the ultimate in "hands-on training courses." This "learn-by-doing" program,

(continued on Page 12)

---

The last of some 280,000 servicemen and women graduated from the former Army Signal Center and School at Fort Monmouth in 1976. (Courtesy US Army Communications-Electronics Command Historical Office, Aberdeen Proving Ground, Maryland) names over the years, to include the Signal School, the Signal Corps School, and the Eastern Signal Corps School. In addition to training scores of Americans for World War I, World War II, Korea, and Vietnam, the school had trained several thousand foreign officers and enlisted men representing some 60 different countries.

It was hard for the local community to accept the departure of the school, which it had championed in so many ways since the World War I era. As James Gannon noted in an oral history interview, "As you know, when we go around and we're speaking at various functions and clubs and groups and businesses in the area, the question I have been asked most, in any event which I have done and I have done quite a bit of it, is 'when is the Signal School coming back to Fort Monmouth?' The Signal School is never coming back. And it is no longer the home of the Signal School. But it is the home of the ... Electronics Command and that's the thing that we've been trying to sell now ..."[11]

## The Equipment Provided

Times certainly were changing. Remember that Brigadier General Lotz quote shared above: "Electronics has never been so vital in a war as it is here in Vietnam." The Electronics Command (ECOM) at Fort Monmouth managed much of those electronics needs—whether that meant innovating the technologies from scratch in the laboratories, adapting technologies produced by private industry, or procuring equipment from private industry partners. What were some of the most important pieces of equipment?[12]

Radios remained a mainstay of the labs, as they had been since World War I. During Vietnam, transistors and integrated circuits replaced tubes. Communications equipment became smaller, lighter, more dependable, and more versatile. It reached lower into the ranks and accommodated a much larger volume than ever before, providing more information to more people, more of the time. Towards this end, General Moorman (commander of ECOM and Fort Monmouth from July 1963–October 1965) ordered the new, transistorized FM radios of the AN/VRC-12/PRC-25 families shipped to Vietnam in July 1965 in response to complaints from General

William Westmoreland (commander of US forces in Vietnam, 1964–1968) about the existing options being too heavy, too unreliable, and too hard to get batteries for.[13] The radios of the AN/VRC-12/PRC-25 families soon became the mainstay of tactical communications in Southeast Asia. ECOM delivered 20,000 VRC-12 and 33,000 PRC-25 radios to Southeast Asia in three and a half years. The PRC-25 was, according to General Creighton Abrams, who commanded military operations in the Vietnam War, 1968–72, "the single most important tactical item in Vietnam."

ECOM also delivered a new radio to replace the AN/PRC-6 in 1967. Troops in Vietnam had found the AN/PRC-6, with its large handset, too awkward for use in combat. The new radio consisted of a helmet-mounted receiver, the AN/PRR-9;[14] and a small transmitter, the AN/PRT-4,[15] which was billed as small enough to fit in your pocket or on your belt. The PRT-4 and PRR-9 worked together as an attempt at a simple, lightweight

AN/PRC-25 radio deployed in Vietnam. (Courtesy US Army Communications-Electronics Command Historical Office, Aberdeen Proving Ground, Maryland)

radio, just for communicating with your squad (versus a longer range radio like the PRC-25). ECOM had their contractors produce sets for 47,000 infantrymen through 1971.[16]

In addition to radios, Fort Monmouth continued to supply troops with various types of radar equipment. The Vietnam War provided the first test of the improved counter-mortar radar AN/MPQ-4 in a tactical environment.[17] The AN/MPQ-4, which had existed in the Army inventory since 1960, was deployed to Vietnam in 1965 and proved particularly useful in the defense of fixed installations.[18] As the *Monmouth Message* boasted in early 1966:

> The MPQ-4 or "one-shot" locator can pinpoint the location of an enemy mortar after only one mortar round comes within range. The round passes through two radar beams and appears as two blips on a radar screen. As soon as the locator operator centers crosshairs on the blips, the map coordinates—or exact location on Army maps—are indicated on a digital readout that looks like the mileage reading on an automobile speedometer. The operator provides the coordinates to friendly artillery or aircraft, and fire can be dropped om the mortar within seconds of its first shot.[19]

Fort Monmouth also developed the AN/PPS-5 man-portable surveillance radar to replace the AN/PPS-4 and AN/TPS-33. The ninety-five pound set had a 360 degree scan capability. It could detect personnel within five kilometers and vehicles within ten. ECOM awarded the production contract in April 1966. There were more than 350 sets in the theater by the end of 1970. Though often unusable for lack of repair parts, the set—when it worked—was popular with the troops because it reduced the need for hazardous surveillance patrols. According to one commander, "One AN/PPS-5 in operating condition is worth 500 men."[20]

The Fort Monmouth labs continued their meteorological work as well. As a 1965 *Monmouth Message* article declared, "The Army's combat forces of the future—with their greatly stepped-up mobility and striking power—will be able to get a faster, closer look at the weather and atmosphere as the result of a new meteorological system being developed under the directions of the Army Electronics Command." The article was referring to the Meteorological Data Sounding System, a "vehicular-mounted, quick-reaction, automated facility to provide sounding data for such purposes as missile firings, conventional artillery, tactical planning, for predicting atomic fallout, and for regular and special weather forecasts."[21] This is but one example of the Fort's continued commitment to understanding the weather as a means of military readiness.

The labs also continued work on satellites, for example the AN/TSC-54. The AN/TSC-54 was a satellite terminal originally designed for quick-reaction use with a military satellite communications, or SATCOM, system. It was completely transportable in two C-130 cargo aircraft. The AN/TSC-54 required only a few hours for a crew of six to manage assembly, installation, and the initiation of operations.[22] Satellites allowed communications over far larger areas than radios.

Night vision became another important Fort Monmouth responsibility during the Vietnam War, building upon the very rudimentary technologies that had existed during World War II. Much of the work was actually conducted at Fort Belvoir, Virginia, but with the Army reorganizations of the 1960s it fell under the supervision of the ECOM labs at Fort Monmouth by 1965.[23] The Small Starlight Scope AN/PVS-2, the Crew Served Weapons Sight AN/TVS-2, and the medium range Night Observation Device AN/TVS-4 all saw service in Southeast Asia. These "revolutionary new devices" enabled US troops "to see in the dark without being seen, thus contributing to military success in Vietnam."[24]

ECOM supplied combat troops with a number of other high-technology commodities during the War to include aerial reconnaissance equipment, surveillance systems, sensors, and air traffic control systems. We have just scratched the surface here. As the Fort Monmouth historian wrote in 1973, "It would be neither possible to list all the accomplishments of the ECOM in its logistical support of the Army in SEA, nor list the extraordinary efforts of the people who contributed in getting the job done."[25] Here, this chapter has attempted to relay the massive scope of the mission and the types of missions accomplished in support of the soldier in the field. Researchers interested in an exhaustive list of devices with technical specifications may enjoy a visit to the US Army Communications-Electronics Command Historical Office archive at Aberdeen Proving Ground, Maryland.

## The Services Provided

In addition to innovating products to meet the demands of soldiers in the field, the Command actively supplied, managed, and supported nearly half the line items in the Army's materiel inventory during the 1960s. (It's one thing to know what you need, but who will make it? In what quantities? How did you transport it to those that need it most? Who is repairing equipment that goes down?) The items ranged in size (from a transistor to a sixty-foot parabolic antenna), complexity (from two-strand twisted wire to airborne surveillance systems), and technologies. The range exceeded that of any other AMC commodity command. Supporting this materiel in the theater involved unique problems and solutions. For example, ECOM struggled to find companies who could deliver quality batteries in sufficient quantity. The command additionally had to worry about how the batteries were stored in the torrid climate of Southeast Asia.

Logistics and supply was truly important work. Rockwell Webb served at Fort Monmouth as a contracting officer during the Vietnam War era, responsible for multimillion dollar procurements. Webb had civilians that worked for him, and their specialty was electronic warfare procurements. An oral history interviewer, trying to understand what it was Webb did, asked "So you would actually be purchasing machine guns, for example?" Webb replied "No. Electronics warfare. Much more—dreamy kind of things." For example, he recalled needing to procure "A radio system that was lightweight and that when the guy was in the jungle, the pilot, they could find him in a chopper" after having to bail out of his aircraft. "Capable to be carried by just one man, and set it up in a big time hurry, and communicate. Because the jungles made it very, very hard to communicate."[26]

ECOM addressed problems of supply and support through a variety of means. Commodity Management Offices (individually dealing with major areas of responsibility like avionics/navigation aids, electronic systems, combat surveillance/night vision/target acquisition, communications/automatic data processing, intelligence materiel, electronic warfare/meteorology, and test equipment/power sources) provided intensive management of critical items. Established when the Command was organized in 1962 and staffed by some of ECOM's best people, the Commodity Management Offices survived in one form or another until 1971.

General William B. Latta (commander of ECOM and Fort Monmouth, October 1965–September 1969) then established a twenty-seven man Operational Readiness Office in 1965. Its sole mission was to monitor the progress and detect the problems of every ECOM project or activity relating to Southeast Asia. ECOM additionally established and staffed the Aviation Electronics Agency and the Avionics Configuration Control Facility in 1966–1967 to address the unique problems associated with installation of ECOM equipment in Army helicopters.

Also in 1965, ECOM also instituted a Direct Exchange/Repair and Return program for nineteen critical items, mostly avionics equipment. Under this program, spares were exchanged for damaged equipment in the theater. Defective components were then returned to the US, usually to the Sacramento Army Depot, for repair and eventual return to the field. As repair requirements changed during the war, so, too, did the number and kind of items on the repair and return list. Defective modules were arriving at Sacramento Army Depot at the rate of 5,000 a month by 1969.

Noting that many modules were damaged or misplaced in shipment, General Latta had the labs design and issue padded, pre-addressed envelopes called "jiffy bags."

The Red Ball Express, instituted by the Army in December 1965, provided emergency supply of critical repair parts and air delivery to Vietnam. ECOM handled 27,000 Red Ball requisitions in 1967, filling 99.2 percent within thirty days (the AMC average during the same period was 97.8 percent). The National Inventory Control Point at ECOM established a permanent office in South Vietnam in January 1968. Civilian supply technicians replaced military expediters to locate equipment in the depots.

Speaking of civilians—ECOM instituted a Technical Assistance Program in Vietnam in 1965 to solve the most troublesome maintenance and support problems on site and also to provide feedback information for correcting design and support deficiencies. One civil servant and thirty-three manufacturer representatives worked the Technical Assistance Program. Latta then organized a formal ECOM Area Office in Vietnam in February 1966. Three years later, the office had a staff of 141 civilian engineers and technicians. Most of the staff was assigned to support MACV Headquarters, the 1st Signal Brigade, the 1st Logistical Command, and the 34th General Support Group.

ECOM also deployed an R&D Technical Liaison Team to Vietnam in January 1967 at the request of the 1st Signal Brigade. The team typically consisted of six or seven people: a team leader and representatives of the R&D Technical Support Activity and the various ECOM laboratories (avionics, electronic components, combat surveillance/target acquisition, night vision, and communications/automatic data processing). Team members typically served three-month tours in theater (leaders, six months) to observe the operation of ECOM equipment, identify deficiencies in design or performance, provide quick-fix solutions, and acquire first-hand knowledge of field conditions. More than eighty ECOM scientists and engineers served on the team between 1967 and 1972. Several served more than once. The team also supported AMC's Vietnam Laboratories Assistance Program. Military and civilian personnel of the ECOM New Equipment Training teams conducted more than eighty missions in direct support of the war in Southeast Asia from 1965 through 1968, including fifty-one missions in theater. More than half of all the missions supported avionics equipment.[27]

That's a lot of different systems and programs—but the main idea with all was to ensure Fort Monmouth based personnel were always ready and accessible to answer the needs of those in the field.

## Soldier Skepticism

The technologies developed, procured, and fielded by Fort Monmouth-headquartered personnel undeniably saved lives. This chapter has included impressive quotes from commanders touting their importance. But it's interesting to note that sometimes soldiers in the field expressed skepticism at the advanced technologies being sent their way. Think about it—when the Wi-Fi works in your house, it's lovely, right? But when something goes wrong with your internet router, you want to cry. Imagine how frustrating—not to mention deadly—equipment malfunctions could be in the heat of battle. For example—for the troops in Vietnam—night vision capability sounds incredible, wouldn't you think? It was—if and when it worked correctly, and everyone knew how to use it! Veteran Rick Amsterdam recalls of the "Starlight Scope" night vision device

> We were told it cost $10k, as if we really cared about the cost. They were not very useful; however, we could make out a silhouette on clear nights. When looking through the Starlight at night, everything looked green, a cross between lime green and Army green, somewhere in between. Silhouettes were black. The silhouette was not distinctive, so we needed to follow-up with illumination flares to identify. It was kind of like there was no way we would put all of our eggs in the Starlight basket. They were heavy and awkward to use ... After a while we went back to using basic listening and visual tactics followed by flares. They were useless in our bunkers because you needed height to get an advantage, hence the reason why we used them in the tower. The guys in the bunker gave us their Starlight due to visual issues from ground level. We ended up with 2 Starlights in the tower that became dust collectors. The reason was the time factor. The time it took to try to identify was far too long to waste. We needed to identify immediately, and the 30 seconds to one minute of time lapse was critical to survival, so we ditched the scopes. Tremendous technology breakthrough for the time, though.[28]

Michael Coale kept his criticism of the Starlight Scope brief, wryly observing "Starlight Scope, just like watching early black and white TV, nothing but static! The only thing missing were rabbit ears to eliminate the static. Ahh the good old days!"[29] Ron Wentworth was similarly unimpressed, noting, "We had one in the tower as well and, to be honest ... I hated the damn thing and stuck to flares!"[30]

Sometimes soldiers yearned for lower tech solutions to their problems. In October 1968, a Signal Corps officer in Saigon wrote to Fort Monmouth to suggest that pigeons be reinstated to help with missions such as delivering requisition requests from units in the field back to headquarters, providing communications for "isolated special forces camps under siege," and "long range recon patrols who must operate under radio silence," and as "an emergency comms method for aircraft on single ship missions who are shot or forced down too quickly to get off a call of their position." The officer wrote, "It occurred to me that carrier pidgeons [sic] with ... messages in aluminum capsules could reduce ... delivery time considerably." He also suggested they might be easier to soldiers to use, and less susceptible to failures or enemy sabotage than higher tech systems, concluding "the thought has occurred to me … that we may be overlooking a very cheap, efficient, and highly reliable way of delivering requisitions" and "I would like to have more information on the capability of pidgeons [sic] in this theater and I feel there is a very possible use for the pidgeon [sic] in the above and other ways here in Viet Nam."[31]

Outlining this skepticism is not meant to detract from the importance of the work done at Fort Monmouth and its satellites, but merely to humanize the experience. The work being done by the research and development, procurement, and logistical magicians of Fort Monmouth only helped soldiers in the field if they understood how to use it, and trusted it. Recognizing this, the Command took great pains in training troops, creating educational materials, and even deploying civilians to help where there could.

## Post-Vietnam

As America's involvement in the Vietnam War ended, the US Army Signal School, a fixture at Fort Monmouth since the World War I, had transitioned to Fort Gordon, Georgia. The post's scientific personnel continued innovating and adapting technologies for use by soldiers in the battlefield as they had since World War I, though Army reorganizations might mean they were working for organizations with different names. A growing Army of contracting and procurement personnel on base worked furiously to try to ensure that the troops had all that they needed. Civilians from Fort Monmouth even deployed in support of the war effort, to provide technical and logistical support on the ground in theater.

ECOM was relatively short-lived. In brief: the Secretary of the Army soon established the Army Materiel Acquisition Review Committee (AMARC) to improve the Army's materiel acquisition procedures. The Committee's report, released in April 1974, said in essence that the commodity command structure of the Army, with its emphasis on "readiness," limited the Army's flexibility and impeded the acquisition process. The Committee recommended that research and development functions be separated from "readiness" functions. Basically, "Project 80" was an experiment that had not been entirely successful.

Troop review, Fort Monmouth, 1969. Following the Vietnam War era and the departure of the Signal School, the number of uniformed military on post steadily declined and Fort Monmouth became overwhelmingly civilian. (Courtesy US Army Communications-Electronics Command Historical Office, Aberdeen Proving Ground, Maryland)

AMARC entailed a two-for-one split for most major subordinate commands of AMC. For ECOM, though, it proposed the establishment of four new organizations: the Communications-Electronics Materiel Readiness Command (CERCOM), the Communications Research and Development Command (CORADCOM), the Electronics Research and Development Command (ERADCOM), and the Avionics Research and Development Activity (AVRADA), a component of the new Aviation Research and Development Command.

But the organizational changes weren't over yet. Eventual reassessment of the AMARC changes concluded that while the emphasis on research and development had increased as desired, there was also much duplication of effort. AMC combined CERCOM and CORADCOM to form the new Communications-Electronics Command (CECOM), effective May 1, 1981. The single command would be responsible for "communications electronics logistic support and material readiness now handled by the

Army communications and electronics material readiness command and research development and acquisition which are now assigned to the Army communications research and development command."[32] Despite the complicated reorganizations, there were relatively few personnel changes and the overarching intent of the work being done at Fort Monmouth remained consistent: the "design and development of sophisticated equipment for the armed forces," which newspapers continued to describe in terms like "electronic wizardry," noting "Although some of the operations performed at the fort are top secret, much of the research that is disclosed is hard to understand because of its highly technical nature."[33]

Other tenants unrelated to communications and electronics missions moved onto the post after the Vietnam War era drew to a close. Though not the focus of this book, these included organizations like the United States Military Academy Preparatory School (which moved to Fort Monmouth from Fort Belvoir, Virginia, in July 1975), and the Army Chaplain Board and School (which moved to Fort Monmouth from Fort Wadsworth on New York's Staten Island in 1979).[34] Prime real estate in central New Jersey could not be wasted! With the land thus put to good use, Fort Monmouth remained open as the country moved into the closing decades of the twentieth century. As an August 1982 newspaper article announced:

> Fort Monmouth is a survivor. From its beginning in 1917 with the establishment of the Signal Corps School to the present, the facility, the largest single employer In Monmouth County, has weathered reductions in forces, the transfer of major commands to other parts of the country and double-digit inflation. Now it is in the midst of the most significant modernization program in the last 29 years, according to Major General Donald M. Babers, commander of the fort. Nearly $25 million in renovations has been completed this year, Babers said. An additional $18 million to $20 million will be spent to upgrade the Hexagon building, which houses many research facilities for the fort ... More than 8,000 civilian and 2,000 military employees work at the fort. The combined military and civilian payrolls totaled $246.6 million in 1981. The fort also has benefitted private business in terms of the number of contracts it issues each year. Babers said the fort awards 30,000 contracts each year totaling $2 billion.[35]

Many organizational changes had occurred but the Fort's commitment to the nation's men and uniform had not. A robust cadre of personnel stood ready to serve the military in Operations Desert Shield and Desert Storm.

CHAPTER 8

# "We Were Able to Control our Force Much Better than the Enemy Controlled His"

## Operations Desert Shield and Desert Storm

The post-Vietnam drawdown meant somewhat reduced missions at Fort Monmouth, but the installation continued its research, development, logistics, and procurement work throughout the latter 1970s and into the 1980s and 90s (figuring out what soldiers needed, and how to get it to them). Despite continued reorganizations on post and the gradual civilianization of the workforce, personnel headquartered out of Fort Monmouth were ready to work round the clock in support of US-led efforts to liberate Kuwait after Saddam Hussein ordered his Iraqi troops to invade his neighbor on August 2, 1990, starting the Gulf War. Organizations based at the Fort would equip soldiers with everything from jammers to night vision, to surveillance and intelligence systems, and to sustain these systems in the field. These systems gave American forces unprecedented capabilities for communication, command and control, surveillance, target acquisition, fire control, position, and data analysis. As usual, some of the equipment was designed on site, some was modified from the civilian world, and some was simply purchased, distributed, and/or maintained by people working in the myriad of organizations and offices that comprised CECOM at Fort Monmouth in the 1980s and 90s.

## The Communications-Electronics Command and the Civilian Majority

As discussed in Chapter 7, the main organization headquartered at Fort Monmouth from 1981 forward was called the Communications-Electronics Command (CECOM). As one colonel, James Gannon, noted, "CECOM is a very large, diverse, complex organization."[1] The Fort Monmouth workforce was largely civilian at this point, which had much to do with the gradual departure of the Signal School during the Vietnam War era (also discussed in Chapter 7). As then Brigadier General George H. Akin, outgoing Deputy to the Commander of the US Army Communications-Electronics Command (CECOM) at Fort Monmouth, noted in 1988, "I think that this particular mission in this Command requires a civilian work force because you've got to have that long continuity. However, I think that I would like to see more 'green suiters' particularly at the field grade and senior non-commissioned officer level, inserted into our various directorates. We are basically building a product for the soldier. If you don't have green-suiters involved, how do we really know what that soldier wants?" Still, Akin concluded, "... the CECOM workforce ... is a tremendous work force in the quality of work and the amount of work we do here ... The mission of this command is so important to the Department of Defense, not just the Army; and our country ..."[2]

The lack of uniformed military on post at this point also gave colonel Raymond Ketchum slight pause, with him noting in an oral history, "I think we're a little short on military."[3] Colonel Gannon concurred, saying:

> Well, of course the mix of civilian has changed constantly since I've been here with less, less military and more civilian. I personally feel that we have, to a great extent, have overall a very dedicated civilian work force. I have been in the service a very long time, worked for many, many years with primarily soldiers, also worked many years with just civilians. I can't say the mix is exactly what I would like to see it because we need to train the military in some of these areas ... I would like to see, in some areas, a few more military. Not because the civilian can't do the job because he can and do it extremely well, but if we want to build military into an organization such as this, how do they get training if we don't have them in it to start so that they can come back to it? In essence, how do we grow the next Commander of CECOM if we don't have military there? Again, not because the civilians can't, because they can do it, extremely well.[4]

While the growing lack of uniformed military personnel on an active duty post like Fort Monmouth was noteworthy, no one doubted that the dedicated civilian workforce was capable of executing the CECOM mission. Any minor concerns about the demographics of the workplace centered on the need to ensure that civilians properly understood what uniformed military needed, and that there were enough uniformed military trained in the workings of CECOM to assume leadership roles at the post as needed.

## Reorganization Redux

Over the past few decades, the Fort had gone from the "Home of the Signal Corps" to the headquarters of the US Army Electronics Command (ECOM). ECOM was then temporarily broken into a few different organizations that were mostly consolidated into the Communications-Electronics Command (CECOM). Still, the Army wasn't done reorganizing. Changes again came to Fort Monmouth with the May 1, 1987 implementation of the Goldwater-Nichols Department of Defense Reorganization Act of 1986. This removed offices known as Project Managers from the Army Materiel Command (AMC) and CECOM control and placed them under something called Program Executive Officers, who reported directly to the Army Acquisition Executive (the Assistant Secretary of the Army for Research, Development, and Acquisition). When asked about this, even seasoned military men were skeptical, with Colonel Ketchum noting:

> We have been trying to shorten the acquisition cycle for years, and I guess that frustration breeds a lot of attempts at organizational changes … The PEO concept, it's fine, it's another organizational change. I believe that whatever it is, as it is developed under Mr. Ambrose and General Wagner and General Bunyard, General Pihl, it will be different when those folks change. And what the difference is, maybe it'll be better, it'll be different, and it will probably be more refined, but I don't know. I've been here long enough, and I've been in the Army long enough to see things go through their full cycle.[5]

The historian interviewing Ketchum, Julius Simchick, observed:

> I spend half of my, better than half, of the inquiry time, from Directors, from the various CECOM organizations, calling me to find out who they were and what they were doing last week. The names have changed so many times.

> The engineer doing his job is sitting at the same desk he always was, working on the same project for the last ten years. He hasn't moved at all, he just wants to know who he's working for because one day he's here, one day he's there, one day he's someplace else, but he's still doing the same thing.[6]

Humorously, Simchick then shares:

> I have [an order] framed behind my desk, from 1983. All it states is BG Morgan's title has been changed from Director of Readiness and Procurement to Director of Procurement and Readiness, period, then the signature. Of all the pieces of paper that came by that want to show change for change's sake, it makes no difference which way the names go, that had to be the top of the bunch. That's why I love that one. That's why it's framed in back of my thing over there. We change too much. We change for the sake of change instead of for the sake of increasing something.[7]

Simchick's view is obviously debatable—some folks with authority clearly thought the seemingly endless organizations at Fort Monmouth made sense. The important thing is that through it all, the faithful Fort employees rarely if ever seem to have faltered in their missions.

## Missions Continue

And those mission were getting more complicated than ever. Former Fort Monmouth command historian Dr. Richard Bingham once wrote, "If a single phrase could be invoked to characterize research and development activities of the 1980s, it would be 'Force Modernization'—the acquisition and fielding of powerful new weapon systems, largely based on technologies developed the previous decade. With automatic data processing systems, such as the Tactical Fire Direction System (TACFIRE), the All Source Analysis System (ASAS), and the Maneuver Control System (MCS), CECOM gave the American Soldier battlefield capabilities no other Army possessed. So did several new surveillance systems, including ... GUARDRAIL ..." This might leave readers asking—what is Dr. Bingham describing?

- The TACFIRE system automated selected field artillery command and control functions to provide efficient management of fire support resources.[8]

- The ASAS was an automated tactical intelligence system that would provide all source correlated intelligence to commanders at division, corps, and echelons above corps. ASAS's all source fusion network could be used to generate timely, accurate, and comprehensive understanding of enemy deployments, capabilities, vulnerabilities, and potential courses of action.[9]
- The MCS, a collection of computer equipment, provided battlefield information by collecting, processing, and displaying data generated within the air/land combat environment. Using this system, a commander could improve the timeliness of his or her decisions and allocate resources accordingly.[10]
- The GUARDRAIL/Common Sensor (GR/CS) was an airborne signals intelligence (SIGINT) collection/location system that provided near real time SIGINT and targeting information to tactical commanders.[11]

During this time, CECOM also introduced new, secure communications systems, including the Single Channel Ground and Airborne Radio Systems (SINCGARS). In 1983, CECOM awarded the first contract for the production of Single Channel Ground and Airborne Radio Systems (SINCGARS) to replace radios of the VRC-12 family. SINCGARS provided Very High Frequency (VHF) Frequency Modulation (FM) combat net radio communication with electronic counter-countermeasures, or frequency hopping, and digital data capability.[12] This meant it had a greater range than older radios, and was harder for the enemy to interfere with. The equipment was also billed as being "tough as a tank."[13]

Another major endeavor of this period was the award of mobile radio-telephone system Mobile Subscriber Equipment (MSE) contracts in a revolutionary $4.5 billion procurement. MSE provided users with a means of communicating throughout the battlefield, regardless of location, in either static or mobile situations. The National Museum of the US Army calls it "the precursor to the modern mobile phone," explaining:

> Mobile or cell networks are so named because calls jump from one zone, or cell, to the next as users move. They can operate completely wirelessly, relying on radio signals rather than hardwired copper cable. They provide data services such as fax or email alongside voice transmissions. Every user, known in tech-speak as a *node*, can

SINCGARS, undated. (Courtesy US Army Communications-Electronics Command Historical Office, Aberdeen Proving Ground, Maryland)

reach every other user via a unique identifying number. All nodes may be literally mobile—callers can move about and stay connected so long as they remain within the coverage area. Researchers with the US Army Communications-Electronics Command (CECOM), headquartered at Fort Monmouth, New Jersey, knew this kind of geographic flexibility was vital to modern battlefield communications. They partnered with commercial contractors to develop a new system optimized for forward operating areas during battle, where permanent equipment was impractical and soldiers were always on the move. Virtually all hardware would be truck-mounted, and all users would be free to move about without losing contact with rearward forces or each other.

MSE was fielded first to the Army Signal Corps, 13th Signal Battalion, 1st Cavalry Division in February 1988. This first-generation MSE system featured phones for both stationary and portable use, fax capability, and channels reserved exclusively for data. Soldiers dialed a fixed number by referring to a directory, similar to a phonebook. MSE software automatically located the desired party anywhere within the network—created by antennas and switching equipment stashed within mobile communications trucks—and connected the call. It required no manual routing or switching. In the event of damaged or overloaded circuits, MSE employed a technique called flood search routing that polled the system for an available circuit and dynamically routed the call around any glitches in the network. MSE also supported now common calling features like call forwarding and early digital or "packet-based" voice communications.[14]

MSE Project Manager Colonel John R. Power, who assumed control just one month after award of the MSE contracts, stated of the program, "I wanted to be a project manager of a major Army Signal program ... I got to be the Project Manager of the Army's most major Signal program."[15]

The Joint Surveillance Target Attack Radar System (JSTARS) evolved from Army and Air Force programs to "develop, detect, locate and attack enemy armor at ranges beyond the forward area of troops." Its primary mission is "to provide theater ground and air commanders with ground surveillance to support attack operations and targeting that contributes to the delay, disruption and destruction of enemy forces."[16] In the simplest possible terms, it's installing radars and other electronics equipment on aircraft to detect targets and relay that information back to command posts on the ground.

As you might have gleaned from these few examples of the inventory of CECOM managed equipment of systems, Army communications were increasingly digital and increasingly complex. Things had come a long way from the wig-wag flags used by the Signal Corps during its founding in the Civil War era, or even the pigeons that had been used in World War I, World War II, and Korea. Even the names of the equipment and systems were getting more complex, with one editorialist musing, "Perhaps no institution has twisted the mother tongue more systematically ... than

Soldiers training with Fort Monmouth systems and equipment, 1995. (Courtesy US Army Communications-Electronics Command Historical Office, Aberdeen Proving Ground, Maryland)

the Defense Department ... The military's seemingly endless supply of acronyms further diminishes the layman's comprehension of defense affairs. The most crucial section of the defense secretary's annual report to Congress, therefore, may be the eight-page appendix of acronyms, which includes ... high-tech jaw-breakers as SINCGARS—V ..."[17]

These "high-tech jawbreakers" would soon be tested on the battlefield.

## Desert Shield and Desert Storm

Busy as it kept in the 1980s, Fort Monmouth was well-positioned to assist when the United States launched air strikes against Iraq on January 17, 1991 in an attempt to liberate Kuwait. CECOM was responsible with equipping and sustaining the force with the communications and electronics equipment it needed to fight. This was not an easy task. Units generally arrived in theater with only the equipment they owned. While some units possessed newer equipment, most units had at least some incomplete or damaged systems. And units often couldn't communicate with each other because their equipment was incompatible. For example, as veteran Bobby Mayfield recalled, "At that time, the Army Guard still had the old style FM radios that had a particular encryption device. The active duty had the SINCGARS. Single Channel—I'm not going to try to remember the acronym, but much more updated radio and they would interact but it—there was no encryption. It was very quirky to try to deal with that."[18] The Army, and CECOM in particular, had to smooth out these types of issues and fill these gaps. CECOM began operating an Emergency Operations Center (EOC) twenty-four hours a day, seven days a week on August 7, 1990 to address the situation. Although several organizations within CECOM set up their own crisis management centers, the CECOM EOC served as CECOM's focal point for all actions relating to the crisis in the Middle East. If soldiers in the field had an issue, CECOM wanted it to be easy for them to find answers.

Employees headquartered at Fort Monmouth worked around the clock in order to equip soldiers with everything from radios and jammers to global positioning systems to night vision and intelligence systems. From day one, CECOM also worked to sustain the equipment out in the field (aka fix what broke) and to ensure any follow-on items arrived in theater mission-ready. For example, the CECOM Readiness Directorate completed 1,318 fieldings between July 1990 and February 1991, many accelerated specifically to meet the requirements for Desert Shield/Desert Storm. In another example, CECOM managed to issue the SINCGARS system to an entire brigade within one week in order to equip the 1st Cavalry Division with SINCGARS radios before its deployment.

This included not only the radios themselves, but also the operator and maintenance support training needed to sustain them. CECOM would repeat this same accomplishment in theater three more times before it was all over. CECOM also supported the war effort through the purchase of commodities: the consumables, repair parts, and replacement items that kept forces viable wherever they operated. This complex, time-consuming process ordinarily involved item managers, contracting officers and other employees across several organizations and functional areas. Much of the usual red tape was temporarily suspended in order to meet the immediate needs of forces overseas.

By the end of the Gulf War crisis, CECOM processed close to 180,000 requisitions, shipped six million pieces of equipment worth over $1.1 billion (including four million batteries), initiated 456 urgent procurement work directives valued at $113 million, and procured a total of 10.8 million pieces of equipment worth $326 million. CECOM also established a Communications Security (COMSEC) Management Office in Saudi Arabia that opened November 15, 1990. While most theaters traditionally had a communications command responsible for managing COMSEC issues, one had not been set up for Operation Desert Shield. The reserve unit ordinarily assigned to Central Command, or CENTCOM, was not deployed due to obsolete equipment. Consequently, the Army Theater COMSEC Management Office (TCMO) was a significant development. CECOM, recognizing the need for dedicated COMSEC support in Saudi Arabia, acquired the necessary authorizations, resources and space to set up at the Royal Saudi Air Force Base in Riyadh. TCMO came under the direct control of CENTCOM shortly after operations commenced and remained operational until May 1991.

CECOM also made extensive use of Logistics Assistance Representatives (LAR) during Desert Shield/Storm. Like their civilian predecessors who had deployed during the Vietnam crisis, LAR were civilian employees from CECOM who deployed to provide hands-on technical assistance when needed. CECOM had forty-eight LAR ready to deploy within seventy-two hours of receiving the full deployment alert for Operation Desert Shield. CECOM LAR proved invaluable in providing assistance whenever soldiers in the field asked for help regarding their equipment.

Government contractors also played a vital role in the Gulf War, as they had in Vietnam. Technical assistance from contractors became necessary in cases involving very recently developed systems on which the effects of the desert (such as the intense heat) were not yet fully understood. CECOM in many cases planned on developing a support capability within the organization but could not do so before the system was sent to the Gulf. In other cases, especially with older items, CECOM no longer had the ability to maintain them. Contractors provided the necessary support. Batteries represented a huge challenge for CECOM during operations. Wartime demands surpassed peacetime stocks. Batteries quickly died, due largely to the intense desert heat. Unfortunately, nearly every item in CECOM's inventory required numerous batteries. Battery producers were instructed to work around the clock by the time the air campaign started in January 1991. This continued until the conclusion of the ground war in early March. Different pieces of equipment, such as radios and night vision devices, demanded different types of batteries. Maintaining stock and ensuring that the right equipment received the right battery became a logistical concern for CECOM. CECOM decided to push shipments into the theater to a single control point for distribution, rather than filling individual requisitions as they were received. The DoD, the Army, and CECOM learned many lessons during Operations Desert Shield and Desert Storm. Although the Gulf War was viewed as an overwhelming success for the nation, the experience demonstrated the undeniable need for enhanced communications and more integration on the battlefield, along with a better logistics infrastructure. These lessons became the impetus that shifted military strategy towards one that emphasized information dominance over brute force.

While there were some lessons to be learned, Fort Monmouth technologies were deployed with great success. Night vision, first introduced in Chapter 7, had become a tried and true technology of this point. 24th Infantry Division Commander Major General Barry McCaffrey commented, "our night vision technology provided us the most dramatic mismatch of the war." Brand new systems like JSTARS quickly won the respect on those in the field, with Brigadier General John Stewart remarking, "JSTARS [Joint Surveillance Target Attack Radar System] was the single most valuable intelligence and target collection system in Desert Storm." A Congressional inquiry further noted, "JSTARS was very successful in identifying large-scale enemy ground movements and helped coordinate attacks against those movements. For example, JSTARS identified the two major routes the Iraqi Army used during their massive retreat from Kuwait City … The JSTARS has been a spectacular success story."[19]

And a March 1991 *Newsweek* article said of the Fort Monmouth-managed Firefinder mortar locating radars' usefulness, "When an Iraqi Battery fired a round, a US Army Q-37 radar would sight it and feed the battery's coordinates to computers that directed the American guns. It took less than a minute to drop a counterround on the Iraqis. Many of them soon stopped firing. To pull the lanyard was to invite death." Firefinder was just the latest in a long string of mortar locating radars Fort Monmouth had supplied dating back decades, each generation modified to meet the needs of those in the field. At the end of the day, one might say that it's testimonials like these that show the true measure of success.

As Dr. Richard Bingham wrote in his 1994 monograph, *CECOM and the War for Kuwait, August 1990–March 1991*, "The roles these systems played in Operation Desert Shield and Operation Desert Storm attest to the significance of the technologies CECOM and predecessor organizations introduced to the battlefield in recent decades and the importance of CECOM support for these technologies during preparations for and pursuit of the Gulf War." The United States and its Allies won the war for Kuwait "not because they amassed a superior force with superior fire power, but because we were able to control our force much better than the enemy controlled his; because we were able to know what the enemy was doing while denying him knowledge of what we were doing; because we were able to see and target enemy forces before they could see us. These were the abilities that [Fort Monmouth]-managed commodities gave the Army."[20]

It would not be long before Fort Monmouth technologies were deployed to the desert again.

## The Command Office Building (COB)

Before getting to that, though, any book chapter about Fort Monmouth in the 1990s would draw the ire of scores of former post employees if it did not mention the closure of the beloved behemoth that was known as the "Command Office Building" or "Green Acres."

This structure was erected at the northwest corner of Tinton Avenue and Wayside Road in Tinton Falls, just a few miles down the road from Fort Monmouth's Main Post and Charles Wood Area. CECOM's predecessor, the Army Electronics Command (ECOM), wanted to consolidate its personnel, many of whom worked in World War II vintage "temporary" structures spread across the Fort's various real estate holdings. After seven years of planning and negotiating, the General Services Administration entered into a binding contract with the Dworman Building Corporation of New York on August 27, 1971. The contract provided for "the construction of a modern six story office building and a 2,500 car parking lot."

The January 25, 1973 *Monmouth Message* boasted that the building would feature thirteen conference rooms, a state of the art auditorium, a self-service supply store, a cafeteria, and a "wide variety of retail stores." There would be "35 vending machines in six separate areas dispensing cigarettes, cigars, candy, soda, and ice cream." There were also designated smoking rooms. As one employee recalled, "You could get lung cancer just from sitting there but the door was kept closed and you could just sit and talk with friends for a few minutes."

The air-conditioned building officially opened on November 19, 1973 when the first 180 ECOM employees moved in. One employee remembered, "People couldn't wait to move there because the building was so new. They started calling it 'The Country Club.'" "Green Acres" initially provided approximately 535,000 square feet of usable office space for an expected 3,500 total personnel. Its enormity could overwhelm. One employee observed, "I remember the parking lots. If you didn't get there early in the morning if felt like a mile hike to get to the building." Another noted, "When I started work with the Army, back in 1979, my first day brought me to the COB. How impressive and intimidating it was. What door do I go in? Do I park in the front or the back? Which elevator do I take?! Then the color coded hallways—green, yellow, blue or orange. I did finally get the knack for finding where I was going. Then I thoroughly enjoyed the shopping and the bar, grill restaurant and the cafeteria. We also had the Credit Union in the building and a mail box." Countless thousands worked there in the ensuing decades.

The thousands of personnel who worked there included some of Fort Monmouth's senior leaders, many of whom remembered the building fondly in their official exit interviews. When in December 2006 outgoing Deputy to the Commanding General Mr. Victor J. Ferlise was asked, "What are some of the things that you've experienced here that would be memorable," his reply included, "when they first closed the CECOM Office Building and moved us on post." Colonel James Costigan, former Military Deputy for Operations and Support, recalled in June 2007 how former CECOM Commander Alfred J. Mallette would go to the "big cafeteria down at the bottom [of the building]. And, you know, he didn't have his lunch brought up. He would stand in line; he wouldn't go to the front of the line. He'd just sit down with people and he would talk [with] them."

When asked in November 2007 if any particular areas [of Fort Monmouth] held a special place in his heart, Mr. Edward Bair, outgoing Program Executive Officer (PEO) for Intelligence, Electronic Warfare, and Sensors (IEW&S), replied, "Area number one is the old CECOM Office Building ... because that's where I began, that's where the roots of my friendships are, and that's where I met my wife." He also laughed, "And it was a god-awful place ... But that's where I began, and that's where my roots are." In fact, Mr. Edward Thomas, Deputy to the Commanding General, CECOM Life Cycle Management Command, has said "I met Eddie Bair standing in line in the lobby of the CECOM Office Building on July 15, 1974, as we were both in-processing on our first day as AMC Comptroller interns." When asked "If you came back ... in twenty years, what would you expect to see, what would you like to see," Bair replied that he would like to come back "twenty years from now and [say], there's the CECOM Office Building, where mom and dad met."

Unfortunately, Mr. Bair will not get that opportunity. The building weathered damage wrought by Hurricane Gloria in 1985, but could not survive a 1993 Department of Defense mandate that CECOM vacate leased space outside of Fort Monmouth and move all activities onto the post. The last personnel vacated the building in December 1998. The structure reverted to the General Services Administration that same month and was then sold into the private sector. After sitting vacant

Top: Few Fort Monmouth buildings captured the workforce's imagination like the CECOM Office Building, 1984. Bottom: The Myer Center in the Charles Wood Area was a close second. The Myer Center, also known as "the Hexagon," was constructed in 1954 to house the Signal Corps Labs. Dedicated in August 1988, it memorializes the founder of the US Army Signal Corps. It was demolished following the base's 2011 closure. (Courtesy US Army Communications-Electronics Command Historical Office, Aberdeen Proving Ground, Maryland)

for many years, the building was demolished 2009 to make way for an adult community—but it lives on it the hearts and memories of many a former Fort Monmouth employee. The COB story illustrates how deeply attached to the individual buildings they worked in personnel could become.[21]

CHAPTER 9

# "It Has Definitely Left Some Lasting Marks that Aren't Necessarily Easy to Get Over"

## Support of the Global War on Terror

At 8:46 on the morning of September 11, 2001, the United States became a nation transformed.

An airliner traveling at hundreds of miles per hour and carrying some 10,000 gallons of jet fuel plowed into the North Tower of the World Trade Center in Lower Manhattan. At 9:03, a second airliner hit the South Tower. Fire and smoke billowed upward. Steel, glass, ash, and bodies fell below. The Twin Towers, where up to 50,000 people worked each day, both collapsed less than 90 minutes later.

At 9:37 that same morning, a third airliner slammed into the western face of the Pentagon. At 10:03, a fourth airliner crashed in a field in southern Pennsylvania. It had been aimed at the United States Capitol or the White House, and was forced down by heroic passengers armed with the knowledge that America was under attack.

More than 2,600 people died at the World Trade Center; 125 died at the Pentagon; 256 died on the four planes. The death toll surpassed that at Pearl Harbor in December 1941.

This immeasurable pain was inflicted by 19 young Arabs acting at the behest of Islamist extremists headquartered in distant Afghanistan. Some had been in the United States for more than a year, mixing with the rest of the population. Though four had training as pilots, most were not well-educated. Most spoke English poorly, some hardly at all. In groups of four or five, carrying with them only small knives, box cutters, and cans of Mace or pepper spray, they had hijacked the four planes and turned them into deadly guided missiles.[1]

The terrorist attacks of September 11, 2001, often simply referred to as "9/11," shook the world. The reverberations were particularly strong in New Jersey, due to the state's proximity to New York's World Trade

Center, or "Ground Zero." One hundred and forty-seven men and women born, raised, or residing at the time in Monmouth County died that day. Many others from Monmouth County were working at Ground Zero that day and survived, forever changed by what they had witnessed. The County's proximity also meant that its first responders and military posts like Fort Monmouth were uniquely situated to assist in the aftermath of the tragedy: for example, Fort Monmouth's firefighters raced to decontaminate survivors as they arrived back to the County covered in toxic dust and debris, and Fort Monmouth's scientific personnel deployed unique equipment to Ground Zero to aid in search and rescue efforts. Unless they specifically asked to be identified, the oral history narrators quoted in this chapter have been anonymized out of an abundance of caution for their privacy and safety, as some are still serving in the military.

## This is Not a Drill

When the September 11 terrorist attacks occurred, the installation was ironically, eerily even, actually in the midst of a three-day force protection exercise involving law enforcement agencies and emergency personnel at all levels, from Fort Monmouth firefighters to the NJ State Police. The exercise included simulating a biochemical terrorist attack at Fort Monmouth and studying the emergency response that would take place after such an attack. As event organizer Mike Ruane recalled in an oral history interview:

> ... the scenario we set up was built on a disciple of Osama bin Laden setting off a chemical release in the Post Theater during an event that the Post was having.
> So we had 75 volunteers. We met at 8:00 o'clock in the morning on 9/11, briefed them as to what their job was. "Okay, you're going to be seated around here. What's going to happen is this supposed chemical release is going to be there. You guys are going to fall in your chairs. And we're going to be videotaping this, okay, because we're going to be using that as part of an overall training package that we're developing."
> So the event takes place supposedly about 8:15 ... we actually had the Fort Monmouth Fire Department and the provost marshals were the first one in there. Next thing you know, we had the Sheriff's Office coming in. Then we had the County Office of Emergency Management. Then we had various other agencies coming in or calling in at that time.

And we had just done our first rehearsal at about 8:30. We finished our first rehearsal. And we're talking to the people and so forth, and I'm in the Post Theater because I was the exercise director. So I'm the puppet master, if you want to call it that. And I'm talking to the people, talking to the videographers, and all of a sudden ... the fire chief at that time, he comes in and tells me, "Hey, we just got something over radio that a plane crashed into the World Trade Center." And I said, "What are you talking about, John?" He said, "No, a plane just crashed into the World Trade Center."

I told everybody, "Okay, everybody just take a break in place right now. Let me—I have to find out something, and I'll be right back in."

So we went outside and while we're out there, the second plane went in. Then we realized that this is what it's going to be. Because the first plane went in, we didn't know the size of the plane or anything else like that. Because planes have crashed into buildings before, we thought it was a small plane ... And when the second plane went in, we knew it wasn't an accident, definitely knew it wasn't an accident. And I went back into the theater, told everybody what had happened, and people actually thought it was part of the exercise.[2]

It took a moment for the employees playacting in this exercise to realize the announcement of the attacks on the World Trade Center were not a part of the drill. As one lieutenant colonel recalled in an oral history interview, "We had to get in front of the people for the exercise ... had to put a halt to the exercise and say, 'What we were just training on is actually happening. Everybody has to go back now to your agency on Fort Monmouth for further instruction.'"[3]

Elsewhere on post, others, too, took a beat to realize the enormity of the situation. It's understandable—not since the Japanese bombed Pearl Harbor on December 7, 1941, had the US homeland been attacked in such a manner. One lieutenant colonel recalled in an oral history interview that he was told, "Hey colonel, listen to this. Some guy flew his plane into one of the towers." He remembered:

I think nothing of it. I go, "Hey, listen, it's probably some guy that had too much to drink. He was a little off his barometer." Just kind of, we, again, military have a ... different sense of humor. Not harmful. Not the folks I trained with anyway ... But we kind of didn't think anything of it ... The thought didn't even enter my mind. And I'm a military policeman. These thoughts always go through my mind. My buddies, my family will tell you that. And it's not paranoia. It's background. It's experience and it's training ... [but on this day] I didn't even think about it [being real].[4]

Very quickly, all post personnel realized the enormity of the situation and sprang into action. One of the first orders of business would be securing the post, which, as this book has discussed, was less than an hour from New York City. As one lieutenant colonel recalled:

> Fort Monmouth ... used to be an open base. You could drive through it any time you want. After 9/11, we closed it off. And we—regular folks, before they could get a force, we were all taking turns standing at the gate. It could have been a civilian with a Department of Defense police officer. It could have been a [West Point] prep school sergeant, with someone else. And we all came together to do that."[5]

The post secured, Fort Monmouth personnel pivoted to supporting rescue and recovery efforts at Ground Zero and the Pentagon in a number of ways.

## Ground Zero

First, Fort fire personnel assisted with survivor decontamination. As Mike Ruane recalled, "Our fire department ended up in the Atlantic Highlands. Okay. When people were coming in from the ferries, they were decontaminating them. Okay, hosing them down."[6] Fort Monmouth fire inspector Tom Braumuller remembered:

> The guys were out at the Highlands where the ferries came across. And they deconned [decontaminated], I think, somewhere around the neighborhood of 5,000 people, to include FD [fire department] and NYPD [New York Police Department]. They were shuttled over. We deconned them. They wanted to know where they were. We said, "you're in Jersey." They said, "We gotta get back." So, went on the ferry and went right back over.

Tom also recalled that Fort fire personnel helped to provide staffing and equipment to New York fire stations devastated that day, and later took shifts sorting through "the pile" (created at Ground Zero by the downing of the towers) in search of human remains.[7]

Bomb disposal specialists from the 754th Ordnance Detachment, a tenant stationed on post, also deployed to Ground Zero to see how they could assist. Engineers and contractor teams, too, deployed to New York, to try to help find survivors in the rubble by locating their cell phones. As one civilian employee on that particular team tells it:

The next morning we were listening to the radio stations out of New York, and they were reporting that people were calling out from the rubble pile, trying to find help, get people to come rescue them. And so my first thought was we have equipment ... spectrum analyzers and things like that, directional antennas that we could use to find some of those people. But the big concern was they would all have somewhat limited battery life on their cell phones. And that's typically no more than 48 hours. So with the knowledge that we could help, and I had an idea of how to do it, I went into the office, my director was there, and my boss was there. He was a division chief ... who has since passed. I told him what my idea was and what I had heard on the radio.

And he said, "Go out in the courtyard and do an experiment." So I went out into the courtyard of our building and got a couple of pieces of equipment and a couple of metal garbage cans, put some cell phones in the garbage cans. And just tried to see if I could see them. We already knew that there was going to be a somewhat heavy metal and reflection of signals situation there, which can make doing the kind of work that we wanted to do very difficult. So I did about 20, 30 minutes worth of experiments out in the courtyard by myself, and I went back in to report the results to my boss. I said, "We can see the signals." So there's a reasonable chance that we could do this. And we can narrow them in within a reasonable percent with the kind of antennas that we had at the time.

So what I didn't know is that while I was out doing this experiment, the leadership from my organization had made contact with the, I believe it was the adjutant general of New—a two star who was at the Javits Convention Center coordinating the National Guard efforts and talked to him about it. And he said, "Yep, get a crew together and get in here." And I went to work that morning not knowing if I would wind up going in, so I didn't have a lot of preparations made for that kind of an experience. So as we were gathering a crew of volunteers to go in and then gathering equipment, I ran home, got some things that I knew I would need, including a camera and my phone, and ran back to the office and started putting everything together in the van that we ultimately took in.

Now, at this point, they had—all of the entrances to the Fort were, with armed guards, and they were all behind buildings that had been heavily sandbagged. So it was a challenge getting on the base for any of us. But there was a crew of, if I remember right, six or seven, including two military from the organization that wound up going in—we were actually delayed 24 hours because what we were told was the FBI didn't want us to come in that early on that day because they were afraid of a building collapse on part of the pile. I guess the hotel that had a big gash in the front of it, that was, I think, in one of the pictures that I sent you, they were afraid that that was going to collapse, and they had to do an assessment of that before they let anybody else come in. So we all went home for the night anyway, realizing that the next morning, bright and early, we would probably be called in to go in ... So [on the 13th] it became just work as fast as you can to get the equipment set up and start getting the initial readings and then start to look for the phones.[8]

Unfortunately, the efforts of this team were hindered by the fact that so many people working on "the pile" had cell phones on them, and there was no way to distinguish between those and the cell phones of survivors who might need assistance. It was further realized, upon seeing Ground Zero, that there were precious few survivors to be saved.

Still, Fort Monmouth-based personnel also used tiny infrared cameras to search through voids in the rubble for signs of life. They brought a laser doppler vibrometer assessed the structural integrity of the buildings, helping to protect rescue workers from building collapses by monitoring the stability of "the pile" and warning of cave-ins. As one civilian analyst

Ground Zero, photographed by a Fort Monmouth employee when deployed in support of search and rescue efforts, September 13, 2001. (Courtesy Melissa Ziobro on behalf of the anonymized employee)

from Fort Monmouth who deployed to Ground Zero recalled in an oral history interview, "I don't think any of us were prepared for what we saw [at Ground Zero] … oh yea, I've got PTSD, any time I hear fire trucks and ambulances … it was very tough going in there, and it has definitely left some lasting marks that aren't necessarily easy to get over."[9]

Fort Monmouth additionally deployed a quick reaction task force to the Pentagon to install a communications infrastructure for thousands of displaced workers there. The installation soon pivoted to support the "Global War on Terror" that followed, and would become the nation's longest war.

## The "Global War on Terror"

It can be difficult to describe, in easily understood terms, how enormous the role of the organizations headquartered at Fort Monmouth, primarily the Communications-Electronics Command, would be in support of Operations in Iraq and Afghanistan. As New Jersey Congressman Frank Pallone explained to his colleagues in Washington, D.C. in October 2001:

> Fort Monmouth is the communications and electronics command for the Army. Anything that involves communications or electronics that is supportive of the war effort against terrorism essentially goes through Fort Monmouth. They do all the research and development under CECOM, Communications and Electronics Command, for the Army, but they are also involved in communications in the field for a soldier that is in place in a theater of war.[10]

A civilian employee of Fort Monmouth who served as the Team C4ISR[11] Senior Command Representative deployed in support of the "Global War on Terror" July–December 2007, noted, "The C4ISR mission is extremely broad in theater. If it deals with communications, computers, intelligence, sensors—you name it—we're probably involved at some point in the life cycle of the system. At the end of the day it's extremely important as our C4ISR systems help save lives and give our Soldiers the decisive edge in battle."[12] And as one colonel on CECOM's leadership team put it:

> generally our role is to act as the single point of contact for C4ISR support across the entire enterprise, across that spectrum for the Army and for the warfighters engaged in Iraq, Afghanistan, the Philippines, and other places. Generally, our roles

run again from fixing whatever is out there, coordinating and orchestrating repair, or sustaining whatever is out there, keeping the adequate number of stockage or repair parts on hand and cached in the right places. This is all so that the entire system can be responsive to the needs of soldiers out in the field. We act as a single point of contract or orchestrator across the spectrum.[13]

What were some of the systems that made a difference? Tried and true systems like SINCGARS and Firefinder and night vision systems were once again deployed. Newer technologies emerged as well. One would be the Common Missile Warning System (CMWS). This system, embedded on aviation platforms, detected incoming heat-seeking and infrared missiles and provided audible and visual warnings to pilots. As an Apache pilot wrote to the command, "I wanted you to know that your product saved my life today. I'm an Apache Longbow pilot deployed to Iraq and while on a mission today I was fired upon. The on-board CMWS deployed and defeated the missile saving myself and my copilot."

Another would be Blue Force Tracking. This system gave troops the ability to track friendly forces on a screen. Today, some readers are surely familiar with technology that allows them to track their family or friends, à la the iPhone's "Find My Friends" function. Blue Force Tracking, cutting edge at the time, helped lower incidents of fratricide.[14] As one colonel recalled:

> There were several instances as far as the initial feedback from people that had gone into combat, where the ability to know instantaneously on the ground and in a very confusing situation that the force you are picking up at sensor range, and at weapon engagement range but not quite in human eye sight range, was friendly, therefore you did not shoot it. Several instances of that were demonstrated by the value of the Blue Force Tracking. That was a big one.[15]

Another colonel concurred, "Warfighters just can't say enough about the benefits Blue Force Tracking provided them on the ground. When they couldn't talk via radio or any other means, they had assured communications through Blue Force Tracking."[16]

The Phraselator also debuted in time for operations in Iraq and Afghanistan. This brick-sized one-way translation device was designed for use by soldiers in countries where they did not know the language and would not have time to learn it. Each handheld unit used an SD

Using Blue Force Tracking, undated. (Courtesy US Army Communications-Electronics Command Historical Office, Aberdeen Proving Ground, Maryland)

card—similar to those used by digital cameras—that stored up to 30,000 common phrases useful for law enforcement, first aid, or war-fighting. To make the device work, a soldier simply said a phrase (such as "stop at this checkpoint") into the device. A few seconds later, the Phraselator would repeat it in the chosen language—for example, Urdu, Arabic, Pashto, or Korean. This system, developed in conjunction with the Defense Advanced Research Projects Agency (DARPA), was critical because there were not enough trained linguists on the ground.[17] As with Blue Force Tracking, you might think—I have apps on my phone that can do this! Keep in mind the first iPhone was not introduced until 2007.

Fort Monmouth based personnel were also responsible for the Lightweight Counter Mortar Radar-Army (LCMR-A), which provided 360 degrees of azimuth coverage and was used to detect, locate, and report hostile locations of enemy indirect firing systems. The LCMR-A was a digitally connected, day/night mortar, cannon, and rocket locating system. It could be broken down, installed in man-packable carry cases, and shipped worldwide without damage by ground, rail, water, and air. Unlike the Firefinder systems, the LCMR did not have a separate search and track beam. Instead, it performed a "track while scan" operation. The Army recognized the original LCMR as one of the Army's "Top Ten Greatest Inventions" of 2004.[18]

Fort-based teams expended much time, effort, and money fighting the threat posed to troops by improvised explosive devices. The Counter-Remote-Controlled Improvised Explosive Device (IED) Electronic Warfare (CREW) Team, including members from throughout Team C4ISR, met an urgent need for Electronic Counter Measure (ECM) devices to defeat the enemy's use of IEDs against coalition forces. One such vitally important device was the WARLOCK ECM test set. The sets protected Army convoys in Iraq, Afghanistan, and other locales in Southwest Asia by detecting and detonating IEDs planted along roadsides. Tens of thousands of these systems were fielded.[19]

## Logistics Assistance Representatives

Because the equipment and systems being deployed were so highly complex and because, as discussed in earlier chapter, CECOM and most of the other organizations at Fort Monmouth were so dominated by civilians, the Fort once again deployed civilians to support efforts in the field. As one civilian noted in an oral history interview:

> The Logistics Assistance Representative program is an Army Materiel Command wide organization. It's not just at the Communications-Electronics Command. There are LARs from ... each one of the commands ... The intent of the program is that there's a realization on the part of the Army that despite the best plans, the best support concept, there's going to be individual times when a unit commander is going to get some equipment and the responsibility for keeping that equipment ready and operational, but something's going to happen that's absolutely outside that commander's capability to influence and get resolved. Maybe the parts aren't available to get in the supply system, or they are broken to the extent that's beyond the level of maintenance that they have in the unit. Any number of

a wide range of problems could occur that that unit commander simply doesn't have organically in his own unit the capability to resolve. The LAR program is out there as kind of a safety net for the Army whereby these Department of the Army civilians (that are stationed at posts, camps and stations and deploy with the units) are able to step in and through their in-depth knowledge of the systems themselves, the technical systems as well as the supply and maintenance systems they support, they know the depots, they know the project managers, they know the weapons system or item manager back at a place like CECOM Headquarters … they know all this and how it's supposed to work and be supported. They're able to intervene and call back and get an entire Firefinder radar antenna, for instance, released from a depot to replace one that might be damaged beyond repair forward. A unit commander wouldn't be able to do that. So the LAR program is out there to be able to step in in these incidents, help the commander, regain the operational readiness of the equipment and then step back again and let the commander continue to operate his unit. So we're not an integral part of the unit, we're not an integral part of the overall supply system; we're kind of like a catch safety net in the event that something happens. So, LARS are normally stationed like I said in places where there's large concentrations of CECOM equipment. There are 41 different offices around the world, all across the United States, Germany, Korea, Italy, Alaska, and the units that they support on a daily basis in garrison deploy, the LARs deploy with them as well.

In Iraq, we started slowly and ramped up to about 85 or 90 people at the maximum, I believe, down now to about 60. The LARs deployed along with the units. When the units went, they went as well. They lived with the units in tents or warehouses, whatever type of facilities were available, and I think probably the common thread of the experience is to travel. It's to be alerted that you're going to go somewhere, Fort Hood as an example, go through the process and get all your shots that you don't already have. In this case, the new one was smallpox, get issued all your gear that you need and be told to report in 24 hours to Building 555 on Fort Hood and be prepared to leave, and inevitably it would be at 3 o'clock in the morning. So you report at 3 o'clock in the morning only to be told that well, there's been a slip, we're not actually going to leave today, come back tomorrow at 3 o'clock. You come back tomorrow at 3 o'clock and find out that there's a change now. Turkey's not letting the ships be off loaded, so we're in a holding pattern, we're not sure where you're going to go now, we thought it was Turkey, but we're still negotiating. Go home but don't leave or go anywhere that you can't be back here in 5 hours notice. So, after some period of time, a couple weeks of preparing, getting ready to go, not going, you might finally board an aircraft and fly from Texas to somewhere like the Azores and then stop in the Azores to refuel and layover for a little bit or maybe the aircraft breaks and has to be repaired. But, eventually after some period of waiting, start going, waiting, going, you finally go, a long journey, 36–48 hours of traveling perhaps half way around the world, you arrive somewhere and you've got 2 or 3 big duffel bags full of stuff, your Kevlar helmets and gear to protect you, and it becomes a scramble then to find transportation. How am I going to get around? Is there going to be a rental car available for me or will I get some kind of 4-wheel drive vehicle? You'll initially be housed in some type of transient billeting place which might just be an open bay warehouse with hundreds of cots.

So, initially there's a transitory period of trying to figure out what you're going to do because you're going to move somewhere forward. Now, you're in another holding pattern while we're waiting to go across the line in Iraq, we're waiting, we're waiting, so now there's another period. So then you move forward and in each case I think there's a period of time involved in uncertainty until you actually move to the next place and every time you move to the next place. Then as you go into Iraq, you move along wherever you're supposed to go. You're given an area to set up and of course it's nothing but racks and partially destroyed buildings and an old warehouse area or an old military compound. You now have to set up tents and create a living space for yourself out of essentially nothing that you've brought with you. And somehow they always manage to land on their feet, and have a good attitude about all the hardships. After everything settles down and it's quiet we get pictures back, for instance some of them that were housed in the palaces, some of them sitting on the old wagons like they came out of 18th century Austrian, some kind of horse-drawn carriage going to the opera in Vienna, this type of thing. So with all this uncertainty, it seems like everyone manages to somehow find the fun part of it, the good part of it, in addition to doing tremendous things as far as supporting the soldiers and the communications electronics systems that they have been helping the soldiers keep the systems out there. They just want to get their job done. That might have been a little more than you wanted to hear, but I get excited talking about it.[20]

Civilians were not forced to deploy; they did so because they so deeply believed in the mission. Supporting the troops had always been personal for the large majority of Fort Monmouth's employees, both civilian and military. When lives were lost in the "Global War on Terror," the team at Fort Monmouth felt it as keenly as the radar folks discussed in Chapter 4 did when the Japanese planes bombed Pearl Harbor. As one senior civilian leader recalled:

> I take it very personally. It is very painful for me to watch the news and see what's happening from two aspects. Number one, it—the personalization of seeing, hearing about, or reading about, or seeing on TV, a young American being killed is extremely personal and painful to me. Particularly, frankly, in some areas that we are responsible for, in terms of force protection, counter-IED devices,

aircraft survivability. And yeah, there's always a "could I have done more? Could we have done more? If we'd only done a little bit more, would we have made a difference? Could we have saved that Soldier's life and everything?" It really strikes home, unfortunately, when you go to a funeral of a young Soldier, and you see their family, and what they're going through at that point in time. Because that could just as easily be my kid. That could just as easily have been either my sons, or it could have been my daughter, or it could have been somebody we know, and certainly we know people whose kids have gone to Iraq. And it's a gut wrenching experience the whole time they're there. And, it is personal to me, and I take it pretty personally, about that.

The second thing, though, about what you say when I look at the news or see that, or read about that, the most frustrating thing about seeing the news and reading the news from my standpoint is, I know, personally, it's not telling the whole story. And that's frustrating to me. Because our kids over there are doing great things in Iraq and Afghanistan, and that's what frustrates me more than

A deployed CECOM civilian uses compressed air to clean a computer printer at Camp Arifjan, Kuwait, undated. (Courtesy US Army Communications-Electronics Command Historical Office, Aberdeen Proving Ground, Maryland)

anything. They don't get enough media attention in terms of, they're building schools, the revitalization or rebuilding of hospitals, the thanks that people actually give them because they have choice in their lives and they're able to do some things that they never before had the opportunity to do or were afraid to do. And it frustrates me that that message doesn't get across, I don't think, as effectively. When I'm sitting in my living room watching the news, all I see is the bad aspects of things, and I don't see the really good stuff that people are accomplishing and the difference they're making.[21]

The growing disconnect between the military and the rest of the country was not lost on the Fort Monmouth workforce. As one colonel noted:

The nation is not at war. The Army is at war, the Marine Corps is at war. We're doing things today that people would get relieved for years ago, because we're trying to do this in such a way, to be more and more efficient. And I think we're doing it because we have to. This is the situation that we have found ourselves in. The young men and women that are in Iraq, Afghanistan, and Korea, all over the world for that matter, the percentage of the overall American population is incredibly small when you look at who is actually doing the fighting. There's more and more Soldiers out there- there's a *Newsweek* magazine, that I have there, and it's called "Fathers, Sons, and War," where, there's more and more Soldiers and their sons and daughters are in the fight. The American public is not in the fight.[22]

It was in the midst of this support to operations in Iraq and Afghanistan that Fort Monmouth would be ordered to close and transfer the bulk of its mission to Aberdeen Proving Ground, Maryland—without disruption to its support of the ongoing wars.

CHAPTER 10

# The Army Goes Rolling Along

Behind the Scenes of a Base Closure[1]

Fort Monmouth was intricately involved in supporting operations in both Iraq and Afghanistan when the Department of Defense (DoD) Base Realignment and Closure (BRAC) Commission convened one of its semi-regular assessments of the military's real estate holdings, with an eye towards where it might divest itself of property to recognize a cost savings or other efficiencies. This chapter gives a behind the scenes look at the ultimately losing battle to save Fort Monmouth from the BRAC mandates of 2005, and explores how and why the Army decided to close Fort Monmouth and shift the bulk of its operations to Aberdeen Proving Ground (APG), Maryland. It wasn't a sign that the missions conducted at Fort Monmouth were unimportant—just that the Army thought they could more resourcefully be conducted elsewhere. This chapter is not a thorough financial audit of the Fort Monmouth BRAC process or an attempt to argue whether or not there were, in the end, any negative impacts on the missions being moved. This author suspects there is a lot of room to play with the numbers as far as the former goes, and that the latter might be impossible to discern. The chapter is, rather, an attempt to memorialize the human impacts of BRAC at Fort Monmouth and the herculean efforts of the employees of the post to execute an unpopular decision to the best of their ability with no playbook. As in earlier chapters, the intent is to allow people to speak to the reader in their own words as often as possible. The oral history narrators quoted in this chapter were all intimately involved in the BRAC process. As in Chapter 9, most narrators have again been anonymized out of an abundance of caution as many still work for the government and their candor might be unappreciated in some corners.

## Past Close Calls

Fort Monmouth had been threatened with severe personnel reductions and even closure several times prior to the twenty-first century. As discussed in Chapter 7, the Signal School had moved from Fort Monmouth to Fort Gordon, Georgia, during the course of the long Vietnam War era. In the mid-1970s, the Army had proposed moving even more of Fort Monmouth's remaining operations to Adelphi, Maryland. A very large and vocal "Save the Fort" campaign ensued. Editorials urged locals to show their support:

> not only to save Fort Monmouth but to save the economy of Central New Jersey. This economy has been geared to the fort and any reduction in its personnel or activity will be dangerously disruptive … there is no question as to the damage [the Army's plan] will inflict on a broad area centered in Monmouth County. The economy of this area was developed to accommodate Fort Monmouth.[2]

Ultimately, far fewer jobs left Fort Monmouth than initially proposed.

Then came the Defense Secretary's Commission on Base Realignment and Closures, which convened in 1988, 1991, 1993, and 1995. The Commission was initially chartered on May 3, 1988 to "recommend military installations within the United States, its commonwealths, territories, and possessions for realignment and closures" which might help the government recognize a cost savings and achieve other efficiencies. The Defense Authorization Amendments and Base Closure and Realignment Act (Public Law 100–526), as enacted October 24, 1988, would provide the basis for implementing the recommendations of the 1988 Commission. Under this Act, all closures and realignments were to be completed no later than September 30, 1995.[3] The specter of Fort Monmouth's closure or severe loss of jobs on post near constantly haunted the Fort during these years. Critics of any Army plans to close Fort Monmouth or significantly reduce the missions headquartered there would again argue, as they had in earlier decades, that "the loss of jobs on and off post will create a tidal wave of economic disaster" and would decrease military readiness.[4] Throughout these years, some missions left Fort Monmouth and some new ones moved onto the post. Some sub-installations of the post, most notably the Camp Evans area in Wall Township, closed but, by and large, the main post escaped the chopping block.[5] Then came the 2005 BRAC round…

The fiscal year 2002 National Defense Authorization Act called for an additional round of base realignment and closure in 2005 by amending the Defense Base Closure and Realignment Act of 1990 (Public Law 101–510).[6] This would be the BRAC Fort Monmouth could not escape. In short, on May 13, 2005, the Department of Defense recommended closing Fort Monmouth and moving the activities located there elsewhere.[7] The outcry was immediate. As US Representative Rush Holt of New Jersey implored his Senate colleagues to consider on May 25:

> We could go through a list of all of the problems that will be created, but let me just paint a picture here. At Fort Monmouth in New Jersey, there are really the best people in the world, mostly civilians, engineers, scientists, procurement specialists, providing communications, surveillance, tracking friendly forces and unfriendly forces, providing equipment, services, software that men and women in the field in Iraq and Afghanistan need and use every day. Thousands of jobs will be sent elsewhere.
>
> Now picture this: A commander in Iraq places an emergency call back to the US. The insurgents have changed the electronics in the roadside bombs, the IED devices, and they need new electronics to detect and disarm them. The reply, "I am sorry, that guy does not work here anymore. We are in the middle of realignment and we have not hired his replacement yet." Repeated 5,000 times, "That guy does not work here anymore," that is what is at stake here.[8]

Despite such pleas, the BRAC Commission approved the DoD's recommendations on August 24, 2005 and issued a report to the president which included closing Fort Monmouth and realigning "Army Team C4ISR"[9] elements at Fort Monmouth to APG and Fort Belvoir, Virginia by September 2011. This recommendation become law on November 9, 2005. At that time, it was expected that the one-time cost to execute the Fort Monmouth closure would be $780.4 million, with an annual recurring cost savings of $146.0 million in years thereafter. There were about 5,000 government personnel (military and civilian) working on post at the time who would be impacted (not to mention local contractors, merchants, and so on and so forth, whose livelihoods were tied to the installation. Fort personnel put the number of jobs tied to the base at close to 23,000).[10]

The government's justification read, in part:

> The closure of Fort Monmouth and relocation of functions that enhance the Army's military value, is consistent with the Army's Force Structure Plan, and maintains adequate surge capabilities. Fort Monmouth is an acquisition and research installation with little capacity to be utilized for other purposes. Military value is enhanced by relocating the research functions to under-utilized and better equipped facilities; by relocating the administrative functions to multipurpose installations with higher military and administrative value; and by co-locating education activities with the schools they support. Utilizing existing space and facilities at the gaining installations maintains both support to the Army Force Structure Plan and capabilities for meeting surge requirements.[11]

Many accepted or even applauded the Commission's work. As one Congressman from New York observed:

CECOM and Fort Monmouth Commander Major General Michael Mazzucchi with Fort Monmouth Garrison Commander Rikki L. Sullivan at a press conference following the announcement that Fort Monmouth would close, 2005. (Courtesy US Army Communications-Electronics Command Historical Office, Aberdeen Proving Ground, Maryland)

This was the least political, most professional BRAC we have ever had. And that is a tribute to Chairman [Anthony] Principi and all of the distinguished members of the panel ... The fact is DoD had too much physical inventory. It is costing DoD to maintain that physical inventory. It is costing the taxpayers. So understandably they wanted some realignment, adjustments; and there had to be winners and losers. As someone who has been on both sides of that issue, let me say I know what it is like. I can feel the pain of the losers. But I would say to those who are on the short end of the recommendation, one, you should have confidence that the recommendations were made once again by the least political, most professional BRAC we have ever had, a BRAC whose individual members, including the Chairman, were available not just to have a courtesy photo opportunity, but to hear out those of us who had presentations before that Commission They asked pertinent questions. They had on-site visits. They were very, very serious about their important work; and they were not alone. The highly dedicated and very competent professional staff of BRAC was even more accessible. You can understand when you get on the phone and you try to get a conversation with Chairman Principi or General So-and-So or Admiral So-and-So, a lot of people want to talk to them. I must say that I was fortunate to be able to talk to each and every one of them. I had quality time. But the fact of the matter is the staff followed through once again with on-site visits, and that was so very important. The dedication and determination demonstrated by the Commission, its accessibility for individual members, their willingness to listen produced a product that I think we can all be proud of.[12]

The praise was not, however, unanimous.

## The Reaction

The BRAC process did not receive such glowing reviews in central New Jersey. Many Fort personnel were flummoxed when the decision to close the post was finally announced. One civilian employee involved in the in-depth fact-finding process prior to the BRAC decision was made recalled:

I participated in an Army Science Board study on the BRAC for [US Army Materiel Command Commander] General [Benjamin S.] Griffin and for Mr. [Francis] Harvey, who was the secretary [of the Army] at the time. And the Army Science Board team that was asked to take a look at the BRAC came to the conclusion that it was a very, very stupid move to close Fort Monmouth and move it to Aberdeen. Especially when they did it on the basis of economy. Because it's going to be years before they recover—not only the money, but the intellectual property they're going to be losing when people start leaving.[13]

Another civilian employee observed:

> ... I'm a business manager who looks at fiscal responsibility and resourcing and all that—to me, it doesn't even make sense what they're doing. I mean, we're spending billions of dollars to move, you know, a whole group of people, and it doesn't—to me, it doesn't make sense from a fiscal point of view at all. I mean, you know, Congress has said we're going to do it. I'm willing to support them, and do that mission, but it doesn't make any sense at all. And I think the work force feels that too. I mean, they look at it and the fact that they don't expect to see any return on the investment for 40 plus years, it's like, you know ...?[14]

One might suggest that these were civilian employees, overly attached to Fort Monmouth, unused to moving, maybe even unwilling to make personal sacrifices, and incapable of seeing the Army's broader plan. But as one colonel noted:

> I would say that I think history will show that there was an opportunity to do something better than moving the Army Team C4ISR to Aberdeen. In the summer of 2005, before BRAC became law, I was in two venues with Major—then, at the time—Major General [Michael] Mazzucchi. In one venue, he told the Chief of Staff of the Army, and in the other venue he told the Secretary of the Army at the time, that moving Army Team C4ISR to Aberdeen made little operational sense. He said that—he said that he wasn't talking about "closing Fort Monmouth," he was talking about "moving Army Team C4ISR." Close Fort Monmouth. If you move Army Team C4ISR to Hanscom Air Force Base [Massachusetts], where the Air Force does C4ISR, there's some operational sense to that, because you get some synergy, overlaps can be taken down, overlaps of capability can be taken down to fill whatever gaps they are. If you move Army Team C4ISR out to San Diego, where the Navy does C4ISR, same situation. If you take all three of those service C4ISR teams, and move them to Offutt Air Force Base [Nebraska] and build a brand new Joint C4ISR facility, that makes operational sense. But to move Army Team C4ISR from Fort Monmouth to Aberdeen—which has absolutely no history of C4ISR—made little operational sense. And I think historically we will find as we look back, the DoD missed an opportunity to move more towards Joint Operations C4ISR by moving it to Aberdeen. We've spent—without a doubt, the Army DoD spent much, much more money building the C4ISR capability campus structure at Aberdeen than it would have if moving to any one of those other places.[15]

This individual was not alone in thinking the Army was perhaps short-sighted and prioritizing the wrong things. As another civilian noted:

It's a business decision by the Army, to consolidate brick and mortar to save money. And that's all it is. It has nothing to do with building an effective mission or an efficient mission. It has nothing to do with taking care of people, it has nothing to do with military value, it has to do with the fact that Fort Monmouth is a tract of 1100 acres that is a highly desirable tract for commercial and residential development, with a great value monetary value, it's in the middle of an affluent community, called Monmouth County, who will recover quite readily, because it is an affluent community, and on the other end of the decision, we have Hartford County, which is in a recession if not a depression, with a high level of unemployment, with a median family income that is far less than Monmouth County family, who are trying to seek job opportunities for its residents. We're moving to an 85,000 acre tract of land called Aberdeen Proving Ground, that by EPA assessment, using that as an indicator, is unsalable without a multi-billion dollar remediation and reclamation activity, to clean up the decades of pollution that have taken place on that Proving Ground. And so I think, at the end of the day, I just take it for what it is. It's a business decision by the Army, to consolidate brick and mortar to save money. And get out of an installation that will recover, and get into an installation that they can't get rid of ... again, you know, the only thing I can say is, to sum up this whole process, it is a business decision by the Army to consolidate brick and mortar to save money. They have yet to prove by their track record that they have ever saved the money they claim they can save, but they will move forward, the mission will move forward, because it has to. Somehow, someone, somewhere, not necessarily at Fort Monmouth or Aberdeen, will fill the voids that will be created by this mission. Unfortunately, there will be risks realized, and now we'll see how good the Army is at managing the risks that they've generated. And hopefully we won't see and come to regret those on the battlefield.[16]

## Concerns about the Mission

Such concerns about the mission were not unique. As one civilian employee noted:

> Well, you know, as I said, I contributed to the report to Congress, we've tried to lay out what needs to be done, none of that is being resourced by the BRAC appropriation. We're going to make it work because we have to make it work; but if we don't make it work, then IEDs [Improvised Explosive Devices] are going to become a prevalent threat again, manpads may become a prevalent threat again, information warfare will become a prevalent threat again. Because these are all activities that are being supported not by some defense industrial base alone, whose building all the solution, but solutions being built right inside

the fence line at Fort Monmouth, and being fought every day. I mean, in an unclassified context, this enemy continues to adapt by coming up with, on average, five different ways to detonate an IED every month. And every month, there's a band of people within this fence line that are working on solutions to defeat those five new ways of detonating an IED. If we don't do this right, I don't know who's going to pick up that work. Because I don't see anyone else in the defense complex stand up and say they're going to do it. They contribute to it, but no one's ever stood up, for all the criticism, and all the bantering, and said, "well, we'll just take that mission on and do it." Ok? So, yes, I have concerns, that if we continue with operations, whether it be Afghanistan, or Iraq, or Central Africa, or anyplace else in the world, we are building risk into the support to those missions. And that is a risk the Army has knowingly and consciously decided to assume by its decision to close and move us. And we'll see where that takes us as time unfolds.[17]

Another civilian wrote:

I am professionally and personally concerned about the ability of our community to successfully relocate the mission of the Team C4ISR to APG without having a negative impact on our support to GWOT and contingency operations. The TF [task force] has laid out a approach, with specific "tools" defined that would enable us to minimize the risk to mission continuity; however, at this time virtually none of those tools have been made available to us by higher headquarters and Congress.[18]

Local politicians were also concerned about the mission, believing firmly in the importance of the work done at the post. As Rush Holt told the press, "We dare not, dare not allow the work at Fort Monmouth to be interrupted. Soldiers in the field depend daily on the work here by the men and women of New Jersey." Then acting NJ Governor Richard Codey called the fort a "center of excellence," enumerating all it contributed to the war effort in Iraq, and lambasting the Pentagon's decision to close the fort as the "wrong idea at the wrong time."[19] New Jersey Senator Frank Lautenberg warned his colleagues in Congress that:

Fort Monmouth, based in New Jersey, is the Army's primary intelligence, surveillance and reconnaissance facility. The Army's work at Fort Monmouth is critical to the safety of America's military men and women and to the success of their missions. The intelligence support it provides goes directly to our troops in the field, making them more effective fighters and protecting their lives and the lives of those around them.

> Over the next 5 years, researchers at Fort Monmouth are slated to develop significant innovations for our Armed Forces, such as Warlock Jammers, which emit radio frequencies that interfere with the signals that set off improvised explosive devices—infamously known as IEDs. The Jammer was engineered at Fort Monmouth and modified for use in Iraq. The military was able to deploy them within 60 days of their development, and they save American lives.
>
> But despite the critical value of this and other innovations at the Fort, the BRAC Commission in 2005 voted to close Fort Monmouth. It goes without saying that no Senator wants to see a base close in his or her State. And it is not only New Jersey that will suffer a loss of jobs and economic activity because of the 2005 BRAC process. But the situation with Fort Monmouth is unique and casts a shadow on the entire base closure process.
>
> As we learn more information about the closure of Fort Monmouth, it becomes increasingly clear that this was a flawed process based on faulty estimates that must be thoroughly investigated. The first and most pressing question is how this closure will affect our troops in the field, given the crucial work Fort Monmouth does for ongoing missions overseas.[20]

Lautenberg's "first and most pressing" question would be easier asked than answered.

## Concerns about the Local Community

In addition to concerns about the impact Fort Monmouth's closure might have on the nation's men and women in uniform, fighting in two wars, there were also understandable concerns about the economic impacts of the closure. While the Army hoped BRAC 2005 would save some $50 billion over the next twenty years by closing dozens of US military facilities, officials in the state of New Jersey, and the county of Monmouth, and the local municipalities were terrified at the thought of losing the area's biggest employer. At the time of the closure announcement, the Fort occupied about 450 acres in Eatontown, about 425 in Oceanport and about 250 in Tinton Falls.[21] Gerry Tarantolo, then mayor of Eatontown, worried:

> Fort Monmouth contributes to the local, county, (and) state economy on the order of $3 billion a year. We have 5,300 jobs directly impacted by this. Possibly another 22,000 jobs that are affiliated with some function at Fort Monmouth. So when you start putting the numbers together, you're starting to approach 30,000 jobs being lost in this area. That's a concern.[22]

Meanwhile, then-Oceanport Mayor Maria Gatta wrote to the BRAC team that "the loss of more than 5,300 people employed by Fort Monmouth due to closure will have a significant impact on the unemployment rate and tax revenues in the region" and that the base closure would "impact Oceanport shops, restaurants, service establishments and businesses," and likely cause "some marginal business to fail, resulting in the loss in tax revenues, subsequently raising property taxes for Oceanport residents and the business community." Further, Gatta pointed out that a sudden flood of vacant office space would affect "the value of office buildings, resulting in a reduction in … tax revenues."[23]

Though his town was going to be one of the most immediately and deeply economically impacted, then-Tinton Falls Mayor Peter Maclearie quickly pivoted to the future, saying of the BRAC decision, "I'm not surprised. Disappointed, but not surprised. Maybe it was all too obvious from the beginning. The lines were drawn. It's time to accept reality and move forward."[24]

In addition to the economic impacts, there were community impacts that were more difficult to measure but emotionally important to many. Fort Monmouth had been a part of the fabric of the community for 88 years when the BRAC decision became law. As one colonel observed:

> … one of the things that BRAC is missing, one of the big negative things about closing this post, is that we're going to lose a footprint of the Army surrounded by great Americans. It's a tiny footprint, I got it, but at least it's a footprint to where folks can drive by and see Fort Monmouth. They can have Colonels and Sergeants Major talk at the different events. It's a way to stay connected with the community. We're going to lose that when we close this post.[25]

No more uniformed soldiers from the base visiting classrooms, Veterans Day and Memorial Day events, and the like. No more open houses welcoming the communities on post to see equipment and demonstrations. No more romances between the soldiers and the locals.

## Concerns about Cost

Remember that a primary goal of BRAC is to recognize a cost-savings for the government. Yet as Senator Lautenberg implored his Congressional colleagues to consider:

... the costs of closing Fort Monmouth are skyrocketing and call into question the very cost-savings rationale upon which BRAC decisions are made. This argument was made by many in 2005, but the warnings were ignored. And as more facts come to light, it becomes apparent that the BRAC Commission was not given all of the information that it should have had to make its decision.

The original cost estimate for closing the fort was $780 million. But according to the Army's own budget request for the fiscal year 2008, that cost has now nearly doubled to $1.5 billion. We all know that the cost overruns are not limited to the closure of Fort Monmouth. In fact, the Congressional Research Service has calculated that overall BRAC costs have increased from initial estimates of $17 billion to a current projection of $32 billion. There are also signs that the true costs of closing Fort Monmouth may have been ignored in 2005. There is mounting evidence that the Pentagon knew, or should have known, that the cost estimates it gave the BRAC Commission related to the closure of Fort Monmouth were not correct. A July 2005 memo from Fort Monmouth officials detailed significant cost errors in the Pentagon's estimates, but the information in that memo was never received by the BRAC Commission.[26]

The costs of closing Fort Monmouth, of required new construction at Aberdeen Proving Ground, and of relocating personnel would continue to be debated for years to come.

## Logistics of a Base Closure

All of these arguments about brain drain and the economic impact to the local communities and the possible impacts to the missions in Iraq and Afghanistan and costs were ultimately considered and discarded. Whether people liked it or not, Fort Monmouth was closing. There would be no reprieve this time around, despite lobbyists' best efforts (and even some lawsuits). How exactly do you close a post down? And let some of the workforce go, and convince some to move to their new duty station? How's it all work?

As one member of the post's BRAC Task Force noted, "We are going to be physically uprooting people from Fort Monmouth, **and** labs, **and** equipment, and we're going to be putting them on trucks and moving them to Aberdeen."[27] As far as the labs and equipment, this was an incredible logistical undertaking. The move involved incredibly delicate, extremely expensive, sometimes one of a kind facilities and systems. In the end, some things would be moved, and much would be rebuilt in the new camps built down at Aberdeen Proving Ground. And as far as the workforce—who would move first? Last? When? And who would make those calls? As one civilian planner noted, "Logistically, you know, when you think about all the people moving, selling their house, moving into a new house, getting kids settled in school, all that, you can't send everybody down at the same time."[28] Everyone recognized what a daunting task lay ahead. The BRAC commission tells you that you have to close, but they don't come and do it for you. According to one BRAC Task Force member, it's "pretty complicated. There's really not a rulebook for how to do it, so you have to kind of invent it as you go along. And that's sometimes good, sometimes not so good when you didn't get it right."[29]

There were simply so many moving parts. As that same individual noted:

> There are a total of eighty-four moving organizations. We broke it down and divided those organizations up. Now, consequently, we've developed a closing checklist for each one of those organizations, and that closing checklist starts at thirty months out, and goes to the very last day that they're here. And it just kind of outlines what they have to accomplish at thirty months, at twenty-nine months, at twenty-four months, at eighteen months. It kind of gives them a road map to closure, which they didn't have before. The other thing that we accomplished is that the ... BRAC Office developed a non-disclosure agreement between the state and federal government. The FMERPA [Fort Monmouth Economic Revitalization Planning Authority], which is the state's local development authority, had asked for maps that were sensitive in nature, and we had to come up with a way to transfer those. So, consequently, we developed this non-disclosure agreement which is pretty unique, and I think that we're the only ones that have that. We've also completed a one hundred percent inventory and provided it to the state. We are now in the process of identifying that inventory as being available for reuse, [or] not available for re-use.[30]

A major concern was whether the workforce would actually move to their new duty stations. As discussed in previous chapters, the Fort Monmouth workforce had become largely civilian over the second half of the twentieth century. There was, early on, a lot of denial about the post closing; a hope it would be overturned. As one civilian employee recalled, "All those articles in the *Asbury Park Press* and all that kind of created this false expectation that it was going to eventually be turned over. And so, it was kind of a hindrance to getting people to accept that BRAC was going to happen."[31] After all, the post had been

THE ARMY GOES ROLLING ALONG • 173

Major General Dennis L. Via at one of many town halls intended to keep Fort Monmouth employees informed as BRAC unfolded, circa 2007. (Courtesy US Army Communications-Electronics Command Historical Office, Aberdeen Proving Ground, Maryland)

threatened with closure or severe reductions many times over the past several decades, and had always come out alright.

At times, there seemed to be a disconnect between some uniformed military leaders on post and their many civilian employees. As one civilian noted:

> The other thing we need to do, quite honestly, is I think the senior leaders—especially the military senior leaders—need to stop looking at this move through the rhetoric of a military officer who's quite accustomed to moving. They need to look through this through the eyes of a civilian, who sits there and says, "This wasn't part of my plan. My plan was to come to Fort Monmouth, serve my country, and retire out of Fort Monmouth." That has changed.[32]

Coming to grips with the post closure took time for many. Three full years after the closure announcement, one colonel noted:

> I think the workforce has **started** to turn to acceptance. Not approval, by far, probably, close to 100 percent remain very vocally disapproving of the decision.

> But I do think that the majority are starting to turn towards acceptance. And acceptance doesn't mean that they're going to go to Maryland, but acceptance means that they're starting to think about their own personal and professional futures with the job that they currently have in Maryland and Fort Monmouth being closed.[33]

It initially looked as if many people would not go, which might create a highly problematic and even deadly "brain drain." In 2007, Senator Lautenberg warned his Congressional colleagues that:

> It is becoming increasingly clear that only about 20 percent of the highly trained and highly skilled workforce who work at the Fort—from engineers to scientists—appear willing to move to Maryland. This is far fewer than the rosy forecast of 75 percent that was provided to the BRAC Commission in 2005. Again, we must ask how this shortage of expertise will affect the critical operations and technology that Fort Monmouth currently provides to our military.[34]

This weighed heavily on the minds of the military and civilian planners at Fort Monmouth tasked with executing the move. When asked in an oral history interview, "Ok, so looking ahead a couple of years from now, what kinds of things concern you?," a civilian BRAC planner noted, "The fact that so many people are not going to move, and how we're going to reconstitute the workforce down at Aberdeen."[35] The task of convincing people to move was truly daunting. One colonel noted in 2007:

> It blows my mind, because, again, knowing what I know about what goes on here, and how they support the organizations in the field—the center of gravity for a commander is information, today. And, we're about to take that center of gravity, and shake it up, and move it. I mean, when we move—when we take an organization and move them from point A to point B, we say, "stand down, get your stuff ready, move, stand back up." We don't have that luxury. We've been told, "close, move to Aberdeen, and then, by the way, don't disrupt operations." When you tell an American GI and his or her family to do that, it's no big deal. Again, thirteen times in twenty-seven years. But when you tell someone here, a civilian who has lived their entire adult lives and raised their children here, and oh, by the way, their spouse doesn't necessarily work for the government, to pack up and move, that's a significant emotional event.[36]

And a civilian leader observed in 2007:

> The last survey, basically says, about twenty-five percent of the people have already made up their mind they're not going. They're either at or will be at retirement and they're not leaving their families, and they're not taking them. Twenty-five percent say they are going, they've already made the decision, they're anxious to get on with it and move, which I think is great, and then there's fifty percent that say they're undecided. Knowing my workforce, I believe the preponderance of the fifty percent is not going. They're just not committing at this point in time to keep their options open and- so, my biggest concern is how quickly we are able to move to Aberdeen and whilst quickly moving to Aberdeen be also recruiting and bringing in a newer workforce at Aberdeen so that we can make the transition. And, in making the transition as expeditiously as possible, we will minimize the impact to the mission. That's my opinion, from my standpoint. I don't really—everybody, to make the move sooner is better from my standpoint than it is prolonging it out to 2011 because I think we run the risk of breaking the mission, the longer that we operate in split based operations, from that standpoint.[37]

The number of employees expected to move to APG did rise as the years crept by, though. Much of it had to do with people just finally accepting the reality of the post closure. According to one colonel in 2009:

> You can see from the BRAC surveys, our percentage of employees that are going to move has grown from 15 percent to about 50 [percent]. Not all of those, of course, are attributed to our retention tools as much as they are [to the fact that] people realize now, after four years, that it's going to happen. And that they are now seriously considering their own personal and professional lives about what is best for them. And a lot of them are coming to the conclusion that [moving to APG] is what is best for them both personally and professionally. Not necessarily what their most desired outcome is, but what is best for them pragmatically is to move down to Aberdeen. So I think we're doing everything that we can. I think also that we've talked before about the strategic opportunity presented by BRAC to transform the organization without having to worry about faces, because so many were still going to attrit. We're still going to have an attrition of 50 percent, which allows the organization an attempt—the opportunities to still have it—to transform, and maybe we don't need as many spaces to do our job as we had. And since we're not going to have to worry about 50 percent of the faces from a personal/personality perspective, we may be able to do something and transform the organization.[38]

Many people found they could no longer retire as they might have hoped, or find a new job in New Jersey, because of the economic downtown of 2007–2008. As one senior military leader said:

I had no idea the economy would change; I'm not an economist, but things change in life. And so naturally, I'm certain this has impacted on some of our employees who had planned to retire. Those who perhaps said, "No, I'm going to retire and so I don't worry about this," [the situation] probably has changed because maybe the opportunities they were looking at may not be as fruitful as they were anticipating. It certainly has had an impact on some of our employees as they look to sell their home. And we're doing everything we can to address that and we'll see what additional information comes out of the [federal government] stimulus [package], and how that will impact as we work with the Corps of Engineers. So I think it has impacted our employees who were already facing a tough decision, and probably made it a little tougher in that regard.[39]

A civilian planner observed that the team responsible for implementing the move needed to handle the changed circumstances of personnel delicately, saying:

The workforce continues to do its job and support the Soldier, in spite of uncertain times. So that's a credit to them. I think we have improved in our information processes, leadership, of getting information out to the workforce and helping them to craft an informed decision. [We need to reach out] not just to employees who were deciding to go, but employees who may decide not to go. I think we're coming to grips with, how do you sell a move? Because part of this is a sales job, to the workforce. I think we need to stop, however, if we really want to continue the positive trend, we need to stop promoting a cost of living argument as a basis or reason to move to Aberdeen. I've been down there; I've done the research. I know others who have done the research. Housing prices are no different than Monmouth County, New Jersey, and they're only getting worse now that they know that there's thousands of families potentially moving in. Baltimore continues to sprawl from the south, and impact property values. Reevaluation cycles are coming up on some towns in Maryland, that will probably increase property tax values above what the claim is, that there's lower property taxes today. I fully anticipate by 2011 property taxes will be probably 70 percent, if not 90 percent, of what they are in Monmouth County. And yet this is all in the midst of the announcement that there will be a four percent decrease in the cost of living allowance to individuals. Maryland has done itself a disservice in coming out often, these last few months, after everything was decided, and saying now they need $15 billion of infrastructure improvements in order to accept the inbound. School systems, roads, hospitals—that's not playing well with employees. Where's that cost of living that I was going to realize, and live better because of? I think the blind spot we have is a quality of life argument. We're not sitting down and drawing quality of life comparisons. Monmouth County, for its high cost of living, has a high quality of life. You can be at the beach in 15 minutes;

you can be in the mountains in an hour. The public school systems, at the primary level, are considered some of the best in the country by the National Department of Education. The same can't be said, unfortunately for Hartford County, in terms of its primary schools. Its secondary schools are comparable to Monmouth County. But we have to pick out those things that are quality of life issues for working in Hartford County and start promoting those in order to keep that positive trend ... the cost of living argument doesn't ring true, when you start to look in those areas, and the quality of life argument, depending on what you're looking to get out of life, may not ring true. So I think if we could just make some minor adjustments, we could be improving that retention rate.[40]

## New Horizons

To be fair, there were folks excited about the relocation and new opportunities it might offer. As one "early mover" from Fort Monmouth to Aberdeen Proving Ground noted in 2009:

> I would characterize us as a lot of people who are eager to move forward with our careers because, I mean, we did make a big change in our lives to really stick with what we're doing, and I think it's beneficial for the organization. I think we're a very strong presence in the organization because we're willing to make our own personal sacrifices to really better the whole BRAC process. But there's a whole lot of new people that I haven't met so hopefully they're just as driven as the early movers are.[41]

Another early mover who felt the move was an opportunity for advancement for emerging professionals in particular recalled:

> I mean, I've had good experiences with it because my fiancé also worked ... for Fort Monmouth and she came down too so ... it really helped us, for where we are, you know, at this stage of our lives, but I would say ... I would think that they would be a little more proactive about it ... I have good things to say about it and I, you know, I've been one of the first people to come so ... it's been a good experience I'd say.[42]

Some early movers felt quite proud about going. When asked, "How do you characterize the C4ISR advanced presence?" One replied:

> They're ... I don't know I hate using the word "trailblazers" but that's basically what we are. I believe these are the people that are the future of Fort Monmouth that, you know, they're coming here. They have the knowledge and the knowhow

of what it was like to relocate. They're going to be taking over the workload from those people that are retiring. So I believe that they're phenomenal in the fact that they made the decision to relocate, to go ahead and come in advance of the mass chaos and hysteria that's going to happen. Not that they're not going to experience more because as those people are hysterical we're going to have to handle the work that they're not doing. But I just think that they're a fantastic group of people to make the decisions they did in coming down here and ensuring that there is a smooth transition, and the Warfighter is not impacted.[43]

Of course, many at Aberdeen Proving Ground and in its surrounding communities were thrilled to have an influx of personnel from Fort Monmouth. A 2007 report by the Maryland Department of Business and Economic Development declared, "The final 2005 Base Realignment and Closure (BRAC) Commission report became law on November 9, 2005, and its recommendations are to be executed no later than September 2011. Maryland fared exceptionally well, gaining an estimated total of more than 45,000 federal and private sector jobs through time, most involving high technology and paying exceptionally well. The BRAC results represent the largest single employment growth activity in Maryland since World War II and will continue to underpin Maryland's movement toward a more stable and increasingly knowledge-based economy."[44]

Major General Randolph P. Strong, the last commander of the Communications-Electronics Command (CECOM) at Fort Monmouth, officially "cased" the CECOM flag at Fort Monmouth in September of 2010, formally signaling the end of the Command's time there. A caretaker workforce, headed by retired Lieutenant Colonel John Edward Occhipinti, would remain behind to ensure the property remained in good repair while vacant and to help facilitate its transition to its next owners via an entity called the Fort Monmouth Economic Revitalization Authority (FMERA).

In the end, the 2005 BRAC impacting Fort Monmouth was completed on time. Subsequently, 4,806 employee positions, 117 laboratories, and 100,000 pieces of equipment and furniture moved from Fort Monmouth to APG. The Army says it was done without interruption to its support of the nation's men and women in uniform, with General Strong observing upon his retirement:

I would say the most significant accomplishment under my command was the undisrupted support to the Warfighter even though we were BRAC-ing … The most significant accomplishment was supporting the Warfighter without a hiccup, without missing a blink, while we were making that move. And you've got to realize that was during a time of not only the surge in Iraq, and drawdown, but also the surge in Afghanistan. So these were very pressing times, demanding times, for the Warfighter. And I think we really, really did that well.[45]

That is a true testament to the tenacity and dedication of the Fort Monmouth leadership, the BRAC planning team on post, and the men and women whose lives were impacted by the move in both positive and negative ways.

The Army spent more than $1 billion on construction at Aberdeen Proving Ground. It added 2.8 million square feet of facilities and eighteen buildings, demolished 140 structures, improved nine miles of roads and upgraded electric, water and information technology infrastructure. As General Strong noted shortly after the move was completed, "CECOM

Groundbreaking for the new Team C4ISR campus at Aberdeen Proving Ground, Maryland, March 2008. (Courtesy US Army Communications-Electronics Command Historical Office, Aberdeen Proving Ground, Maryland)

really gained by this decision. This is a world-class facility that we did not have at Fort Monmouth. This relocation really postures the command well for the future in order to be responsive to the Army's needs and to be relevant to the Army of tomorrow."[46] The new campus incorporates some memorials and markers from Fort Monmouth as a way of facilitating esprit de corps and commemorating the ninety-four years of magic that occurred there, in central New Jersey, on behalf of the nation.

The old Fort Monmouth land, meanwhile, moves slowly into the next phase of its life. Officers' quarters have been turned into gorgeous private homes that retain their historic character. The old athletic center is now a state of the art private gym called "The Fort Athletic Club," which incorporates military themes throughout. As this book is being written, there are plans for streaming giant Netflix to establish a large footprint on site. Just as some memorials and markers moved to APG, some remain behind to pay homage to the men and women, military and civilian, who served our country there from 1917 to 2011.[47] Sidenote: There have been no subsequent BRAC rounds. As Aaron Mehta wrote for *Defense News*, "There were BRAC rounds in 1988, 1991, 1993 and 1995, but the most notorious is the 2005 edition. While the Pentagon says it is now enjoying the benefits of that effort, the 2005 drill was seen as excessively costly and left members of Congress, already sensitive to political ramifications of bases leaving various states, with a built-in aversion to further efforts."[48]

# Conclusion

This book has, as often as possible, told the story of Fort Monmouth's history through the words of those who lived it. It will close now with some words from General (Retired) Dennis Via, the next to last commander of CECOM at Fort Monmouth (who, incidentally, would go on to become the only commissioned Signal Corps officer in the US Army's history to achieve the rank of four-star general). As Via noted in 2009:

> There are nine decades that people have worked here; there are nine decades of stories, technological innovations. Think of the thousands of people who've come from here. This has been the home of the Signal Center; this has been the Women's Army Corps; intel was here. This was the former Chief of Chaplain's office, and that's why I sometimes joke that I feel doubly blessed because I'm sitting here in the former office of the Chief of Chaplains. But this is a historic place ... nine decades, we ought to celebrate that ... Yes, BRAC is here and we're going through—but there is a lot to be proud about, there's a lot to reflect upon. People's parents have worked here, grandparents, cousins, aunts, uncles, mother, father. And I routinely run into people who say, "Yeah, you know, I came here in 1930-something, or 1940, or 1950." I run into them in the PX or commissary, and those are special people. You know, they were here, and a lot of what we have today, and the capabilities we have, were born out of Fort Monmouth ... this great base, this historic base.[1]

"This great base, this historic base," contributed to the defense of the United States and of democracy around the globe for over ninety years. Its training, and its technologies, saved lives on the battlefield. And the legacy of "this great base, this historic base," is everywhere—if you know where to look. It's deeply imbedded in military history and the current operations of the military, to be sure. Innumerable pieces of equipment being used by all branches of the military, and really militaries around the world, were developed or perfected by Fort Monmouth-based teams. But you can also think of Fort Monmouth the next time you listen to FM radio in your car, or check the weather forecast, or pick up your cell phone, or send an email, or fly in an airplane, or spot a pigeon flying past, or see a satellite blinking overhead, or get a ticket thanks to radar clocking you speeding (ok, maybe don't think of Fort Monmouth then). The point is, the smart and tenacious men and women, military and civilian, who worked at Fort Monmouth 1917–2011 changed the very world that we live in. The talented and generous local communities helped them do it. And no post closure can take that proud heritage away from the region.

# Endnotes

## Preface and Acknowledgements

1. In Rebecca Klang, "A Brief History of Fort Monmouth Radio Laboratories." Unpublished manuscript. Courtesy US Army Communications-Electronics Command Historical Office, Aberdeen Proving Ground, Maryland.
2. Fort Monmouth employee interviewed by Wendy Rejan, May 30, 2008. Courtesy US Army Communications-Electronics Command Historical Office, Aberdeen Proving Ground, Maryland.
3. Jerry Dean Swinehart interviewed by Ralph Lane, April 17, 2002. Courtesy Library of Congress Veterans History Project, Washington, DC.
4. "Colonel Cowan Leaves and Colonel Helms Arrives at Camp Alfred Vail," and "Training Battalion Notes," *Dots and Dashes*, July 10, 1918.
5. Clarence Norton, Jr. interviewed by Bridget Federspiel and Toree Wilson, May 8, 2008. Courtesy Library of Congress Veterans History Project, Washington, DC.
6. Eldon C. Hall, "From the Farm to Pioneering with Digital Control Computers: An Autobiography," *IEEE Annals of the History of Computing* (April–June 2000), 24.
7. *Final Report, USASRDL Centennial Presentation* (Fort Monmouth, NJ: United States Army Signal Research and Development Laboratory, 1960), 179. Courtesy US Army Communications-Electronics Command Historical Office, Aberdeen Proving Ground, Maryland.
8. George H. Akin interviewed by William Strong, April 4, 1988. Courtesy US Army Communications-Electronics Command Historical Office, Aberdeen Proving Ground, Maryland.
9. Albert James Myer became the first and only Signal Officer in 1860. It was not until 1863 that Congress authorized a regular Signal Corps for the duration of the War.
10. Raymond E. B. Ketchum II interviewed by William Strong, July 10, 1987. Courtesy US Army Communications-Electronics Command Historical Office, Aberdeen Proving Ground, Maryland.

## Introduction

1. Sidney Shallett, "The Magicians of Fort Monmouth, NJ," *Saturday Evening Post*, August 23, 1952.

2   "Monmouth to Stage Special Observance," *Asbury Park Sunday Press*, March 1, 1953.
3   "New Jersey's Revolutionary War Sites: Site & Visitor Readiness Assessment," *NJ.gov*, 2020-site-assessment-executive-summary.pdf. (nj.gov).
4   Personal letter from Major General Charles H. Corlett to Colonel Sidney S. Davis dated December 3, 1955. Courtesy US Army Communications-Electronics Command Historical Office, Aberdeen Proving Ground, Maryland.
5   Ibid.
6   Historical Branch, G3, "History of Fort Monmouth 1917–1953." Courtesy US Army Communications-Electronics Command Historical Office, Aberdeen Proving Ground, Maryland.
7   John Marchetti interviewed by Fred Carl, January 9, 1999. Courtesy US Army Communications-Electronics Command Historical Office, Aberdeen Proving Ground, Maryland.
8   George Raynor Thompson and Dixie R. Harris, *The Signal Corps: The Outcome* (Washington, DC: Office of the Chief of Military History, 1966), 15.
9   Harold Zahl, *Electronics Away or Tales of a Government Scientist* (New York: Vantage Press, 1969), 120–125.
10  Gloria Stravelli, Army research facility once known as the 'Black Brain Center,'" *Greater Media*, March 27, 2003, https://archive.centraljersey.com/2003/03/27/forts-gate-swung-open-to-black-scientists/. Accessed May 18, 2023.
11  *Electronics*, Volume 39, Issue 2, 1966.
12  Fort Monmouth employee interviewed by Melissa Ziobro, July 29, 2021. Author's collection.

## Chapter 1 The Racetrack Paves the Way

1   For more, see Jean Soderlund, *Lenape Country Delaware Valley Society Before William Penn* (University of Pennsylvania Press, 2014) and *Separate Paths: Lenapes and Colonists in West New Jersey* (Rutgers University Press, 2022).
2   "New Jersey's Revolutionary War Sites: Site & Visitor Readiness Assessment," *NJ.gov*, 2020-site-assessment-executive-summary.pdf. (nj.gov). Accessed March 30, 2023.
3   For more, see Maxine Lurie and Richard Veit, *New Jersey: A History of the Garden State* (Rutgers University Press, 2012) and Maxine Lurie, *Taking Sides in Revolutionary New Jersey: Caught in the Crossfire* (Rutgers University Press, 2022).
4   Michael Strauss, "Monmouth Park Posts a Centennial," *New York Times*, July 19, 1970; Linda Doughtery, *The Golden Age of New Jersey Horse Racing* (CreateSpace Independent Publishing Platform, 2016); Sharon Hazard, *Long Branch in the Golden Age: Tales of Fascinating and Famous People* (The History Press, 2007); Rick Geffken and Muriel J. Smith, *Hidden History of Monmouth County* (The History Press, 2019).
5   Melissa Kozlowski (Ziobro), "Fort Monmouth: Home to the Jersey Derby?," *Monmouth Message*, March 22, 2005; Melissa Ziobro, "Monmouth Park: A Summer Attraction for Centuries," *Two River Times*, June 8, 2023.

6   Lynn Rakos, *Cultural Resources Investigation, Poplar Brook Flood Control Feasibility Study* (New York: US Army Corps of Engineers, 1999), E-5.
7   *History of Fort Monmouth, 1917–1953* (Fort Monmouth: Signal Corps Center, 1953), 3.
8   "At The Summer Resorts," *New York Times*, July 1, 1870.
9   Howard Green, *Words That Make New Jersey History* (New Brunswick: Rutgers University Press, 1995), 154.
10  Robert Russell and Richard Youmans, *Down the Jersey Shore* (New Brunswick: Rutgers University Press, 1993), 35; Evaluation of Selected Cultural Resources at Fort Monmouth, New Jersey, 1995.
11  Russell and Youmans, *Down the Jersey Shore*, 35.
12  "Monmouth Park," *New York Times*, July 9, 1874.
13  "At The Summer Resorts," *New York Times*, July 1, 1870.
14  Ibid.; Historical Office Staff, *Fort Monmouth, New Jersey: A Concise History* (Fort Monmouth: Communications-Electronics Command, various dates); John T. Cunningham, *This is New Jersey* (New Brunswick: Rutgers University Press, 1978), 223.
15  "Business At The Sea-Side; How The Summer Hotels Are Flourishing," *New York Times*, July 17, 1881.
16  Newmarket, in Britain, was the site of England's first organized horserace. Centuries old, the site was frequented by royalty and boasted world class thoroughbreds. "Monmouth's New Course; A Race Track Modeled After Newmarket Heath," *New York Times*, January 21, 1889; *History of Fort Monmouth*, 3.
17  "Racing At The Branch: Inauguration of the New Monmouth Park Race-Course," *New York Times*, July 31, 1870; Historical Office Staff, *Fort Monmouth, New Jersey*; George H. Moss Jr. and Karen L. Schnitzspahn, *Victorian Summers at the Grand Hotels of Long Branch, New Jersey* (Sea Bright: Ploughshare Press, 2000), v, 21.
18  Racetracks in New Jersey were not a new concept. While generally meeting the approval of the localities where they were located, tracks in Jersey City and Camden closed in 1845 after the public complained about "bookmakers, hoodlums, and drunken crowds." Moss and Schnitzspahn, *Victorian Summers*, 21; John T. Cunningham, *New Jersey: A Mirror on America* (Andover: Afton Publishing Company, 1997), 158.
19  Lawrence Galton and Harold J. Wheelock, *A History of Fort Monmouth, New Jersey, 1917–1946* (Fort Monmouth: Signal Corps Publication Agency, 1946), 12.
20  "Racing At The Branch; Inauguration of the New Monmouth Park Race-Course." *New York Times*, July 31, 1870; Michael Strauss, "Monmouth Park Posts a Centennial," *New York Times*, July 19, 1970; Moss and Schnitzspahn, *Victorian Summers*, 21, 35; George H. Moss Jr., *Twice Told Tales: Reflections of Monmouth County's Past* (Sea Bright: Ploughshare Press, 2002), 35, 46; "Historical Properties Report, Fort Monmouth, New Jersey and Sub-installations Charles Wood Area and Evans Area," July 1983. Courtesy US Army Communications-Electronics Command Historical Office, Aberdeen Proving Ground, Maryland.

21 Historical Office Staff, *Fort Monmouth, New Jersey.*
22 Moss and Schnitzspahn, *Victorian Summers*, 21; Helen C. Pike and Glenn D. Vogel, *Eatontown and Fort Monmouth* (Great Britain: Arcadia Publishing, 1995), 81.
23 Bingham, *Fort Monmouth, New Jersey;* Galton and Wheelock, *History of Fort Monmouth,* 12; "Historical Properties Report, Fort Monmouth, New Jersey and Sub-installations Charles Wood Area and Evans Area," July 1983. Courtesy US Army Communications-Electronics Command Historical Office, Aberdeen Proving Ground, Maryland.
24 "Monmouth Park," *New York Times,* July 9, 1874.
25 Jerome Park opened in 1866 near Fordham, New York. It was the headquarters of the American Jockey Club.
"Monmouth Park," *New York Times,* July 9, 1874; "At The Summer Resorts," *New York Times,* July 1, 1870.
26 Moss and Schnitzspahn, *Victorian Summers,* 21; Moss, *Twice Told Tales,* 46; Russell, *Down,* 35.
27 "At the Summer Resorts," *New York Times,* July 1, 1870; Moss and Schnitzspahn, *Victorian Summers,* 21, 26; Moss, *Twice Told Tales,* 46; Sarah Allaback, et al, *Resorts and Recreation: An Historic Theme Study of the NJ Coastal Heritage Trail Route* (Mauricetown: The Sandy Hook Foundation and the National Park Service, 1996), 13; Galton and Wheelock, *History of Fort Monmouth,* 12.
28 "Racing At The Branch; Inauguration of the New Monmouth Park Race-Course," *New York Times,* July 31, 1870.
29 "Long Branch Races," *New York Times,* August 2, 1871.
30 Moss and Schnitzspahn, *Victorian Summers,* 26.
31 "Monmouth Park," *New York Times,* June 29, 1872.
32 "Summer At The Springs," *New York Times,* July 15, 1873.
33 Moss and Schnitzspahn, *Victorian Summers,* vi, 24
34 "Monmouth Park," *New York Times,* June 29, 1872.
35 "Women Who Bet On Races," *New York Times,* August 24, 1885.
36 Ibid.
37 Ibid.
38 "Monmouth Park Races," *New York Times,* July 5, 1878; "Prospects at Long Branch: The Season of 1882 Promising to be Brilliant," *New York Times,* April 2, 1882; "Kinney Beats Pontiac," *New York Times,* July 5, 1885; Moss and Schnitzspahn, *Victorian Summers,* 26; Cunningham, *This is New Jersey,* 223.
39 "Race Track Memories Linger Here," *The Signal Corps Message,* 4 January 1946; Bingham, *Fort Monmouth, New Jersey;* Moss and Schnitzspahn, *Victorian Summers,* 26; "An Archeological Overview and Management Plan for Fort Monmouth (Main Post), Camp Charles Wood and the Evans Area," 1984.
40 "Horses And Horsemen," *New York Times,* January 12, 1890.
41 *Fort Monmouth Yearbook* (New York: Yearbooks Publishing Company, 1947); Allaback, *Resorts,* 14; Cunningham, *This is New Jersey,* 223.

42 Moss, *Twice Told Tales*, 47; Russell, *Down*, 55–56.
43 "Freddie Gebhardt and His Horse Eole," *Northern Pacific Farmer*, October 4, 1883.
44 Ibid.
45 "Twenty-six Years Ago They 'Got the Message Thru'," *The Signal Corps Message*, 9 July 1943; Galton and Wheelock, *History of Fort Monmouth*, 12.
46 Russell, *Down*, 56.
47 Moss and Schnitzspahn, *Victorian Summers*, vi.
48 "Monmouth Park," *New York Times*, July 9, 1874.
49 Pike and Vogel, *Eatontown*, 112.
50 Galton and Wheelock, *History of Fort Monmouth*, 12.
51 "An Archeological Overview and Management Plan for Fort Monmouth (Main Post), Camp Charles Wood and the Evans Area," 1984.
52 "Three Stake Races Run," *New York Times*, June 28, 1891.
53 "Want Racing At Monmouth," *New York Times*, January 25, 1892; "Monmouth Park Closed; Red Bank And Long Branch Feel Very Badly About It," *New York Times*; March 24, 1891; "Racing News And Notions; Little Chance For Much More Racing In New-Jersey," *New York Times*, March 23, 1891.
54 This is eerily reminiscent of concerns over the closing of Fort Monmouth discussed in Chapter 10.
55 "Monmouth Park Opened," *New York Times*, July 5, 1892.
56 "Tracy Bronson's Injunction," *The Jersey City News*, October 23, 1893; "The End of Monmouth Park: Both the Old and the New Track to be Sold at Auction in April—History of the Courses," *New York Times*, March 14, 1897; "Legislative Bribery: Laws for the Protection of Gamblers Secured at Trenton," *New York Times*, October 22, 1893.
57 "Warning To Decent People; Bad Associations Encountered At Monmouth Park," *New York Times*, July 16, 1893.
58 "Scenes at Monmouth Park: Fall of a Noted Racetrack," *New York Tribune*, July 19, 1893.
59 Untitled article, *The Penn's Grove Record*, September 1, 1893.
60 "Murder at Monmouth," *The Monmouth Inquirer,* August 10, 1893; "Donovan sentenced," *The Monmouth Inquirer,* November 23, 1893; "Snip Donovan's Plea," *The Jersey City News,* November 16, 1893; Untitled article, *The Monmouth Inquirer,* April 11, 1895; "Noted Turfman Dead," *The Evening Times*, October 21, 1902.
61 John A. Harnes, "Fort to Celebrate 90 Years of Army Service," *Asbury Park Press*, March 19, 1992.
62 "Uniting Against the Gamblers: Republicans and Democrats Combine in Opposition to Monmouth Park," *New York Daily Tribune*, October 10, 1893.
63 "Monmouth Park Denounced; Press and Pulpit Score Its Methods," *New York Daily Tribune*, July 30, 1893.
64 "Race-Track Politicians," *New York Times*, October 27, 1893.

65 "Legislative Bribery: Laws for the Protection of Gamblers Secured at Trenton," *New York Times*, October 22, 1893; Cunningham, *New Jersey*, 256; Thomas Fleming, *New Jersey: A History* (New York: Norton and Co., Inc., 1977), 146–147; Moss and Schnitzspahn, *Victorian Summers*, 28; Russell, *Down*, 55; Pike and Vogel, *Eatontown*, 83.
66 "New Jersey Legislature; Two Race-Pool Bills Presented In The Assembly Last Night," *New York Times*, February 9, 1892.
67 "The Race Track Denounced; War Against the Reopening of Monmouth Park," *New York Times*, April 4, 1892.
68 "Denounced from the Pulpit; The Rev. Mr. Young's Attack on Monmouth Park Racing," *New York Times*, July 4, 1892.
69 Untitled manuscript, US Army Communications-Electronics Command Historical Office, Aberdeen Proving Ground, Maryland; Galton and Wheelock, *History of Fort Monmouth*, 12.
70 "Betting Kings Doomed," *New York Daily Tribune*, September 23, 1894.
71 "New Jersey Racing Ended; Owners of Tracks Satisfied that the Decision Is Effective," *New York Times*, January 9, 1894.
72 "Monmouth Track Sold," *New York Times*, March 22, 1895.
73 Cunningham, *New Jersey*, 256; Fleming, *New Jersey: A History*, 146–147; Moss and Schnitzspahn, *Victorian Summers*, 28; Russell, *Down*, 55; Pike and Vogel, *Eatontown*, 83.
74 "An Archeological Overview and Management Plan for Fort Monmouth (Main Post), Camp Charles Wood and the Evans Area," 1984.
75 Galton and Wheelock, *History of Fort Monmouth*, 12.
76 Indeed, betting on horse races in New Jersey would not be legalized until 1939. The current Monmouth Park Racetrack, located in Oceanport, was not opened until 1946. "The End of Monmouth Park: Both the Old and the New Track to be Sold at Auction in April—History of the Courses," *New York Times*, March 14, 1897.
77 "Hotel at Monmouth Park is Destroyed," *Asbury Park Evening Press*, April 7, 1915; "Thirty Years Ago: 1915," *Asbury Park Evening Press*, April 1, 1945; *This is Fort Monmouth* (Fort Monmouth: 1950); Moss and Schnitzspahn, *Victorian Summers*, 28; Pike and Vogel, *Eatontown*, 112.

## Chapter 2 "A Big Farm for Soldiers": The Great War Comes to Central New Jersey

1 B. H. Liddell Hart, *History of the First World War* (London: Pan Books, 1972), 1.
2 Christopher Clark, *The Sleepwalkers: How Europe Went to War in 1914* (New York: Harper Perennial, 2014); Council on Foreign Relations, "Why Did World War I Happen?," https://world101.cfr.org/contemporary-history/world-war/why-did-world-war-i-happen.
3 Russell Weigley, *History of the United States Army* (New York: Macmillan Publishing Company, Incorporated, 1967), 377.

4   Historical Branch, G3, "History of Fort Monmouth, 1917–1953." Unpublished manuscript. Courtesy US Army Communications-Electronics Command Historical Office, Aberdeen Proving Ground, Maryland.
5   US Army Center of Military History, "World War I Fact Sheet," https://history.army.mil/html/bookshelves/resmat/wwi/_documents/WWI_Fact_Sheet.pdf.
6   For a thorough history of the Signal Corps, see Rebecca Robbins Raines, *Getting the Message Through: A Branch History of the US Army Signal Corps* (Washington, DC: US Army Center of Military History, 1996).
7   "Signal Corps Recruiting," *New York Times*, June 17, 1917.
8   Lawrence Galton and Harold J. Wheelock, *A History of Fort Monmouth, New Jersey, 1917–1946* (Fort Monmouth: Signal Corps Publication Agency, 1946), 15; "Historical Properties Report, Fort Monmouth, New Jersey and Sub-installations Charles Wood Area and Evans Area," July 1983. Courtesy US Army Communications-Electronics Command Historical Office, Aberdeen Proving Ground, Maryland.
9   Personal letter from Major General Charles H. Corlett to Colonel Sidney S. Davis dated December 3, 1955.
10  Ibid.; "Mr. Van Keuren Purchases the Old Linden Park Race Track; Monmouth Park Too," *Jersey City News*, December 4, 1902.
11  Historical Office Staff, "Fort Monmouth, New Jersey: A Concise History." Communications-Electronics Command, December 2002; Pike and Vogel, *Eatontown*, 80; "Passing of Monmouth Park; Once Famous Race Course of Jersey Cut Up Into Building Lots," *New York Times* (1857–Current file), April 10, 1910; An Archeological Overview and Management Plan for Fort Monmouth (Main Post), Camp Charles Wood and the Evans Area, 1984; "Huge Fort Monmouth Integral Part of Shore Economy," *Asbury Park Evening Press*, November 17, 1954; Stenographic record of interview with Colonel Carl F. Hartmann, Signal Corps Retired, October 26, 1955 in the Office of the Chief Signal Officer; An Archeological Overview and Management Plan for Fort Monmouth (Main Post), Camp Charles Wood and the Evans Area, 1984; Historical Properties Report, Fort Monmouth, New Jersey and Sub installations Charles Wood Area and Evans Area, July 1983.
12  Lawrence Galton and Harold J. Wheelock, *A History of Fort Monmouth, New Jersey, 1917–1946* (Fort Monmouth: Signal Corps Publication Agency, 1946), 12.
13  John A. Harnes, "Fort to Celebrate 90 Years of Army Service," *Asbury Park Press*, March 19, 1992.
14  Memo, The Landmarks and Memorials Subcommittee, Fort Monmouth Tradition Committee, September 21, 1961. Courtesy US Army Communications-Electronics Command Historical Office, Aberdeen Proving Ground, Maryland. See also Edward L. Walsh, "Fort Monmouth: Surviving Four Wars," *Asbury Park Press,* May 4, 1980; "Fort Will Dedicate Col. Hartmann Gate," *Asbury Park Evening Press*, June 13, 1962; Historical Branch, G3, "History of Fort Monmouth, 1917–1953." Unpublished manuscript. Courtesy US Army

Communications-Electronics Command Historical Office, Aberdeen Proving Ground, Maryland.
15 Stenographic record of interview with Colonel Carl F. Hartmann, Signal Corps Retired, October 26, 1955 in the Office of the Chief Signal Officer; Untitled manuscript, US Army Communications-Electronics Command Historical Office, Aberdeen Proving Ground, Maryland; *History of Fort Monmouth,* 5; "Fort Will Dedicate Colonel Hartmann Gate," *Asbury Park Evening Press,* June 13, 1962; John A. Harnes, "Fort to Celebrate 90 Years of Army Service," *Asbury Park Press,* March 19, 1992; "Fort Monmouth Began Modestly as Just a Tent City," *Asbury Park Press,* August 1, 1982.
16 General Order Number 60, "Designation of Hartmann Gate," June 18, 1962. Courtesy US Army Communications-Electronics Command Historical Office, Aberdeen Proving Ground, Maryland. In the interest of space, this book will offer biographies of only certain key figures in Fort Monmouth history. Those interested in additional brief biographies of top-level leadership should see *A History of the Commanding Officers of Fort Monmouth, New Jersey and the US Army CECOM Life Cycle Management Command* (2007), locally produced by and available via the US Army Communications-Electronics Command Historical Office, Aberdeen Proving Ground, Maryland. As for Hartmann, he would depart on July 12, 1917, to establish a similar Signal Corps Camp at Fort Leavenworth, before heading overseas to set up a Signal Corps school for American Expeditionary Forces in France. He was promoted to Colonel, Regular Army, July 20, 1920, with rank from July 1, 1920. He was retired from active service September 27, 1920, and died July 8, 1961.
17 *Dots and Dashes* was the camp's own newspaper. "The Story of Camp Vail," *Dots and Dashes,* June 26, 1918; see also Rebecca Klang, "A Brief History of Fort Monmouth Radio Laboratories." Unpublished manuscript. Courtesy US Army Communications-Electronics Command Historical Office, Aberdeen Proving Ground, Maryland.
18 Historical Branch, G3, "History of Fort Monmouth, 1917–1953." Unpublished manuscript. Courtesy US Army Communications-Electronics Command Historical Office, Aberdeen Proving Ground, Maryland. See also "Sunday Chatter," *Asbury Park Sunday Press,* March 20, 1960; John A. Harnes, "Fort to Celebrate 90 Years of Army Service," *Asbury Park Press,* March 19, 1992.
19 "Shore Veteran Recalls Early Days of Post," *Monmouth Message,* May 18, 1967, 14; see also Rebecca Klang, "A Brief History of Fort Monmouth Radio Laboratories." Unpublished manuscript. Courtesy US Army Communications-Electronics Command Historical Office, Aberdeen Proving Ground, Maryland.
20 Galton and Wheelock, *History of Fort Monmouth,* 16.
21 Historical Branch, G3, "History of Fort Monmouth, 1917–1953." Unpublished manuscript. Courtesy US Army Communications-Electronics Command Historical Office, Aberdeen Proving Ground, Maryland.

22  Laura Mondt, "'Ma Bell,' the Signal Corps and the Great War," East Illinois University, https://www.eiu.edu/historia/Mondt2013SE.pdf.
23  "In the Signal Service," *The Pioneer Express*, September 28, 1917.
24  "Fort Monmouth to Celebrate Two Birthdays," *Asbury Park Evening Press*, June 20, 1958; "Monmouth to Stage Special Observance," *Asbury Park Sunday Press*, March 1, 1953; "Huge Fort Monmouth Integral Part of Shore Economy," *Asbury Park Evening Press*, November 17, 1954; Edward L. Walsh, "Fort Monmouth: Surviving Four Wars," *Asbury Park Press*, May 4, 1980.
25  "The Story of Camp Vail," *Dots and Dashes*, June 26, 1918.
26  For more on AT&T's role in World War I, see Rosanne Welch and Peg A. Lamphier, editors, *Technical Innovation in American History: An Encyclopedia of Science and Technology* (New York: Bloomsbury, 2019).
27  "Signal Corps Recruiting," *New York Times*, June 17, 1917.
28  Historical Branch, G3, "History of Fort Monmouth, 1917–1953." Unpublished manuscript. Courtesy US Army Communications-Electronics Command Historical Office, Aberdeen Proving Ground, Maryland.
29  Rebecca Klang, "A Brief History of Fort Monmouth Radio Laboratories." Unpublished manuscript. Courtesy US Army Communications-Electronics Command Historical Office, Aberdeen Proving Ground, Maryland.
30  Historical Branch, G3, "History of Fort Monmouth, 1917–1953." Unpublished manuscript. Courtesy US Army Communications-Electronics Command Historical Office, Aberdeen Proving Ground, Maryland.
31  Centers for Disease Control and Prevention, "1918 Pandemic," https://www.cdc.gov/flu/pandemic-resources/1918-pandemic-h1n1.html.
32  "History of Camp Alfred Vail 1918," Unpublished manuscript. Courtesy US Army Communications-Electronics Command Historical Office, Aberdeen Proving Ground, Maryland.
33  "Died," *Philadelphia Inquirer*, April 20, 1918; "Three Battalions At Vail Under Quarantine," *Asbury Park Evening Press*, September 27, 1918; "Private W. D. Marks," *Asbury Park Evening Press*, October 16, 1918; "Lieutenant Harold Tierney,"*Asbury Park Evening Press*, October 24, 1918; Galton and Wheelock, *History of Fort Monmouth*, 20; "History of Camp Alfred Vail 1918," Unpublished manuscript. Courtesy US Army Communications-Electronics Command Historical Office, Aberdeen Proving Ground, Maryland.
34  William Blair, "Army Radio in Peace and War," *The Annals of the American Academy of Political and Social Science*, March 1929, Volume 142, 86–89.
35  "Radio School is Begun At Vail," *Asbury Park Evening Press*, December 8, 1917.
36  Lawrence Galton and Harold J. Wheelock, *A History of Fort Monmouth, New Jersey, 1917–1946* (Fort Monmouth: Signal Corps Publication Agency, 1946), 12; Max L. Marshall, *The Story of the US Army Signal Corps* (New York: Franklin Watts, Inc., 1965). See also *The History of Fort Monmouth, 1917–1953*, prepared by the Historical Branch, G3, Fort Monmouth, New Jersey, 1953 and Rebecca Klang,

"A Brief History of Fort Monmouth Radio Laboratories," unpublished manuscripts courtesy US Army Communications-Electronics Command Historical Office, Aberdeen Proving Ground, Maryland.

37 For more on the Signal Corps and early flight, see Rebecca Robbins Raines, *Getting the Message Through: A Branch History of the US Army Signal Corps* (Washington, DC: US Army Center of Military History, 1996) and Dwight R. Messimer, *An Incipient Mutiny: The Story of the US Army Signal Corps Pilot Revolt* (Sterling, Virginia: Potomac Books, 2020).

38 Galton and Wheelock, *History of Fort Monmouth*, 20; "To Enlarge Camp Vail Flying Field," *Asbury Park Evening Press*, July 9, 1918.

39 Galton and Wheelock, *History of Fort Monmouth*, 20–21.

40 "Lietu. L. J. Merkel Killed," *New York Times*, August 17, 1918; "Aviator Killed in Fall at Camp Vail," *Asbury Park Evening Press*, August 12, 1918.

41 Galton and Wheelock, *History of Fort Monmouth*, 23–24.

42 William Blair, "Army Radio in Peace and War," *The Annals of the American Academy of Political and Social Science*, March 1929, Volume 142, 86–89.

43 Kathy R. Coker and Carol E. Stokes, *A Concise History of the US Army Signal Corps* (Fort Gordon, Georgia: US Army Signal Center, 1991), 19. Cher Ami returned to the United States and died at Fort Monmouth in 1919 as a result of his wounds. For more about the Lost Battalion, see Thomas M. Johnson and Fletcher Pratt, *The Lost Battalion* (Nebraska: Bison Books, 2000).

44 Historical Branch, G3, "History of Fort Monmouth, 1917–1953." Unpublished manuscript. Courtesy US Army Communications-Electronics Command Historical Office, Aberdeen Proving Ground, Maryland.

45 "Homing Pigeons," transcript of a lecture delivered by Ray R. Delhauer, Camp Alfred Vail, 1922. Courtesy US Army Communications-Electronics Command Historical Office, Aberdeen Proving Ground, Maryland.

46 Historical Branch, G3, "History of Fort Monmouth, 1917–1953." Unpublished manuscript. Courtesy US Army Communications-Electronics Command Historical Office, Aberdeen Proving Ground, Maryland.

47 Ibid.

48 *History of Fort Monmouth*, 6.

49 Historical Branch, G3, "History of Fort Monmouth, 1917–1953." Unpublished manuscript. Courtesy US Army Communications-Electronics Command Historical Office, Aberdeen Proving Ground, Maryland.

50 Galton and Wheelock, *History of Fort Monmouth*, 16–17.

51 "At Monmouth Park," *Red Bank Register*, July 18, 1917.

52 "Sunday Baseball for Army Benefit," *Asbury Park Evening Press*, June 28, 1918.

53 "Newspapers on File at the Y," *Dots and Dashes*, June 10, 1918.

54 "English Class for Beginners," *Dots and Dashes*, June 19, 1918.

55 "History of Camp Alfred Vail 1918," Unpublished manuscript. Courtesy US Army Communications-Electronics Command Historical Office, Aberdeen

Proving Ground, Maryland. See also Rebecca Klang, "A Brief History of Fort Monmouth Radio Laboratories." Unpublished manuscript. Courtesy US Army Communications-Electronics Command Historical Office, Aberdeen Proving Ground, Maryland.

56 "Local Pastors Sunday Subjects," *Asbury Park Evening Press*, November 17, 1917; "Asbury Entertained 12 Camp Vail Soldiers," *Asbury Park Evening Press*, November 30, 1917; Historical Branch, G3, "History of Fort Monmouth, 1917–1953." Unpublished manuscript. Courtesy US Army Communications-Electronics Command Historical Office, Aberdeen Proving Ground, Maryland.

57 "Red Bank," *Asbury Park Evening Press*, September 25, 1917.

58 "Notes from the 13th Service Co.," *Dots and Dashes*, July 17, 1918.

59 "Big Farewell, Dance," *Asbury Park Evening Press*, March 5, 1918; "Farewell Dance," *Asbury Park Evening Press*, September 26, 1918; "Will Show Many Jumpers: Record Entry List at Hollywood This Week," *New York Times*, July 22, 1918.

60 "Local Pastors Sunday Subjects," *Asbury Park Evening Press*, November 17, 1917; "Asbury Entertained 12 Camp Vail Soldiers," *Asbury Park Evening Press*, November 30, 1917; "Soldiers His Guest," *Asbury Park Evening Press*, December 26, 1917; Historical Branch, G3, "History of Fort Monmouth, 1917–1953." Unpublished manuscript. Courtesy US Army Communications-Electronics Command Historical Office, Aberdeen Proving Ground, Maryland.

61 "Soldiers Are Welcomed by Long Branch Folk," *The Sun*, July 7, 1918.

62 "Sunday Baseball for Army Benefit," *Asbury Park Evening Press*, June 28, 1918.

63 "Letters from the Boys in the Service," *The Fargo Forum*, November 6, 1917.

64 "The Story of Camp Vail," *Dots and Dashes*, June 26, 1918.

65 Ibid.; "Observes Anniversary," *Asbury Park Evening Press*, April 10, 1918; "Liberty Day Celebration at Y," *Dots and Dashes*, May 1, 1918; John A. Harnes, "New from the Front," *Asbury Park Press*, November 10, 1998.

66 "Announce Engagements," *Asbury Park Evening Press*, December 27, 1917; "Whitehurst-Ayres," *Asbury Park Evening Press*, January 25, 1918; "Red Bank," *Asbury Park Evening Press*, October 5, 1928; "Entertainment at Bridge," *Asbury Park Evening Press*, February 1, 1930.

67 "Signal Corps Member Weds Red Bank Girl," *Asbury Park Evening Press*, October 22, 1917; "Cupid is Active in County Seat," *Asbury Park Evening Press*, June 30, 1919; "Robertson-Allen," *The Birmingham Age-Herald,* June 30, 1918.

68 "Red Bank Weddings," *Asbury Park Evening Press*, January 10, 1918.

69 *Dots and Dashes*, May 1, 1918.

70 "Await Ruling on Camp Liquor Ban," *Asbury Park Evening Press*, September 28, 1917; "Selling Liquor To Signal Corps In Uniform," *Asbury Park Evening Press*, September 29, 1917.

71 "Hold Booze Seller," *Asbury Park Evening Press*, October 16, 1917.

72 "Charge He Bought Beer for Soldiers," *Asbury Park Evening Press*, October 22, 1917.

73 "Fort Monmouth Locale Steeped in American Military Tradition," *Monmouth Message*, May 18, 1967; "Fort Monmouth Observes Its 50th Anniversary," *Asbury Park Evening Press*, May 14, 1967; "History of Camp Alfred Vail 1918," Unpublished manuscript. Courtesy US Army Communications-Electronics Command Historical Office, Aberdeen Proving Ground, Maryland.
74 "News of Army and Navy," *Washington Post*, May 18, 1919.
75 "The Story of Camp Vail," *Dots and Dashes*, June 26, 1918.
76 "Tells His Experience After Entering Service," *Evening Star*, October 1, 1917.
77 "Colonel Cowan Leaves and Colonel Helms Arrives at Camp Alfred Vail," and "Training Battalion Notes," *Dots and Dashes*, July 10, 1918.
78 "Training Battalion Notes," *Dots and Dashes*, July 17, 1918.
79 "Survey of Real Estate Owned or Controlled by War Department," *Hearings Before a Special Subcommittee and the Committee on Military Affairs, House of Representatives, 67th Congress, January 12, 1921–April 6, 1922, Part 2* (Washington, DC: US Government Printing Office: 1927), 166.

## Chapter 3 The Inter-War Years: Radar and Other Research and Development Revelations

1 "Camp Vail May Be Permanent Post," *Asbury Park Evening Press*, February 27, 1919.
2 For more on the Army in the inter-war years, see William Gardner Bell, et al, *American Military History* (Washington, DC: US Army Center of Military History, 1989), Chapter 19, "Between World Wars."
3 "Camp Vail's Future Still Undetermined," *Asbury Park Evening Press*, September 6, 1919.
4 "110,000 for the Purchase of Vail," *Asbury Park Evening Press*, November 10, 1919.
5 See George Helms biography file and the records of the memorialization and tradition committee, both available via the US Army Communications-Electronics Command Historical Office, Aberdeen Proving Ground, Maryland.
6 "School to Hold Last Graduation," *Monmouth Message*, June 16, 1976.
7 "Asks $1,500,000 Camp Vail School," *New York Times*, May 25, 1920.
8 "Survey of Real Estate Owned or Controlled by War Department," *Hearings Before a Special Subcommittee and the Committee on Military Affairs, House of Representatives, 67th Congress, January 12, 1921–April 6, 1922, Part 2* (Washington, DC: US Government Printing Office: 1927), 166.
9 "Pershing at Camp Vail," *New York Times*, August 10, 1924; "Pershing to Pay Camp Vail Visit," *Asbury Park Evening Press*, August 8, 1924.
10 "Geran Will Urge Vail Be Created Permanent Camp," *Asbury Park Evening Press*, September 6, 1923.
11 "Camp Vail Made Permanent Camp, With the Name of Fort Monmouth," *Asbury Park Evening Press*, August 13, 1925.
12 "Camp Vail Made Permanent Camp, With the Name of Fort Monmouth," *Asbury Park Evening Press*, August 13, 1925.

13 "Camp Vail Renamed Fort Monmouth; Made Permanent," *New York Herald*, August 24, 1925.
14 "Camp Alfred Vail Now Fort Monmouth," *Washington Post*, August 16, 1925.
15 William Blair, "Army Radio in Peace and War," *The Annals of the American Academy of Political and Social Science*, March 1929, Volume 142, 86–89.
16 "US Seeks Experts: Civil Service Commission Announces Vacancies in Radio Laboratories," *Philadelphia Inquirer*, December 9, 1923.
17 SCR stands for Set, Complete Radio, though it is often interpreted as Signal Corps Radio. Lawrence Galton and Harold J. Wheelock, *A History of Fort Monmouth, New Jersey, 1917–1946* (Fort Monmouth: Signal Corps Publication Agency, 1946) and *The History of Fort Monmouth, 1917–1953*, prepared by the Historical Branch, G3, Fort Monmouth, New Jersey, 1953, unpublished manuscript courtesy US Army Communications-Electronics Command Historical Office, Aberdeen Proving Ground, Maryland.
18 William Gardner Reed, "Military Meteorology," *Geographical Review*, Vol. 12, No. 3 (July 1922), 403.
19 Lawrence Galton and Harold J. Wheelock, *A History of Fort Monmouth, New Jersey, 1917–1946* (Fort Monmouth: Signal Corps Publication Agency, 1946) and *The History of Fort Monmouth, 1917–1953*, prepared by the Historical Branch, G3, Fort Monmouth, New Jersey, 1953, unpublished manuscript courtesy US Army Communications-Electronics Command Historical Office, Aberdeen Proving Ground, Maryland.
20 William R. Stevenson, *Miniaturization and Microminiaturization of Army Communications-Electronics, 1946–1964* (Fort Monmouth: US Army Electronics Command, 1966).
21 "Army's Mystery Ray," *New York American*, August 3, 1935; "Mystery Ray Finds Ships at Sea in Dark," *New York Times*, July 31, 1935; and "Mystery Ray 'Sees' Enemy at 50 Miles," *New York Times*, July 30, 1935; "Army's New Mystery Eye Sights Ships Off Highlands," *Asbury Park Evening Press*, July 30, 1935.
22 William H. Baumgartner interviewed by Daniel Packer, October 31, 2002. Courtesy Library of Congress Veterans History Project, Washington, DC.
23 "New Radio Beam to Guard US Coasts From Air Attack," *Philadelphia Inquirer*, June 29, 1941.
24 Kathy R. Coker and Carol E. Stokes, *A Concise History Of The US Army Signal Corps* (Fort Gordon, Georgia: US Army Signal Center, 1991), 20; Max L. Marshall, *The Story of the US Army Signal Corps* (New York: Franklin Watts, Inc., 1965); Roger Colton, "Radar in the United States Army: History and Early Development at Signal Corps Laboratories, Fort Monmouth, NJ," *Proceedings of the Institute of Radio Engineers*, No. 331, November 1945.
25 Melissa Kozlowski (Ziobro), "Fort Monmouth Officer Inducted into NJ Inventors Hall of Fame," *Monmouth Message*, October 7, 2004; Melissa Ziobro, "William R. Blair," *On Point: The Journal of Army History*, Volume 22, Number 2 (Fall 2016), 19–22.

26  Melissa Ziobro, "Fort Monmouth's Signal Corps Scientists Feared Hindenburg Connection," *Monmouth Message*, May 21, 2010.
27  For more on Camp Evans, see Melissa Ziobro, "Camp Evans, NJ," *On Point: The Journal of Army History*, Volume 26, Issue 2 (Winter 2021), 45–48.
28  "Molly Pitcher," *New York Times*, August 11, 1930; "Many Events Marked 1930 Shore History," *Asbury Park Evening Press*, December 31, 1930; "Hawk in Army Pigeon Lofts Kills Prize Bird, Fights Man," *Asbury Park Evening Press*, February 3, 1932.
29  Paul W. Kearney, "Feathered Soldiers," *Cosmopolitan* (New York) Volume 111, Issue 3, (Sep 1941): 70.
30  Earl Pope Cook interviewed by Lynda Dent, August 4, 2004. Courtesy Library of Congress Veterans History Project, Washington, DC.
31  George H. Manning, "Senate Passes Fort Monmouth Construction Bill," *Asbury Park Evening Press*, April 29, 1926.
32  "Ask $290,000 for Shore Army Post," *Asbury Park Evening Press*, March 1, 1932; "Army Housing Program Must Await Federal Fund Surplus," *Asbury Park Evening Press*, July 30, 1932. For a complete accounting of when buildings at Fort Monmouth were built and for whom they were named, see *Landmarks, Memorials, Buildings and Street Names of Fort Monmouth, New Jersey and the US Army CECOM Life Cycle Management Command*, prepared by the staff of the US Army Communications-Electronics Life Cycle Management Command, July 2009. Courtesy US Army Communications-Electronics Command Historical Office, Aberdeen Proving Ground, Maryland.
33  Melissa Ziobro, "Fort Monmouth Summer Camp Trained Citizen Soldiers," *Monmouth Message*, August 30, 2007.
34  "'Starving' Soldiers Fed: Signal Corps Pigeon Brings Help to Hungry Monmouth Students," *New York Times*, August 28, 1928.
35  For more on the CMTC, see Donald M. Kington, *Forgotten Summers: The Story of the Citizens' Military Training Camps, 1921–1940* (San Francisco, California: Two Decades Publishing, 1995).
36  "Civilian Hospitality Keeps Spirit High at Monmouth and Hancock," *Herald Tribune*, August 3, 1941.
37  "Two Girls Don Soldier Suits, Stroll Into Camp," *Herald Tribune*, July 27, 1941.

## Chapter 4 "Should They Fail, Expect Plenty of Hell": Training and Equipment Critical to Winning The Good War

1  Melissa Ziobro, "Fort Monmouth Invention Warned of Pearl Harbor Attack," *Monmouth Message*, September 21, 2007.
2  John Joseph "Jack" Slattery interviewed by Mark Slattery, June 3, 2001. Courtesy US Army Communications-Electronics Command Historical Office, Aberdeen Proving Ground, Maryland.
3  John Marchetti interviewed by Fred Carl and James A. Broderick, January 9, 1999. Courtesy US Army Communications-Electronics Command Historical Office, Aberdeen Proving Ground, Maryland.

4   John Joseph "Jack" Slattery interviewed by Mark Slattery, June 3, 2001. Courtesy US Army Communications-Electronics Command Historical Office, Aberdeen Proving Ground, Maryland.
5   Harold Zahl, *Radar Spelled Backwards* (New York: Vantage Press, 1972), 73–76.
6   Melissa Ziobro, "Fort Monmouth Invention Warned of Pearl Harbor Attack," *Monmouth Message*, September 21, 2007.
7   William R. Stevenson, *Miniaturization and Micro-miniaturization of Army Communications-Electronics 1946–1964* (Fort Monmouth: US Army Electronics Command, 1966); Max L. Marshall, *The Story of the US Army Signal Corps* (New York: Franklin Watts, Inc., 1965); Roger Colton, "Radar in the United States Army: History and Early Development at Signal Corps Laboratories, Fort Monmouth, NJ," *Proceedings of the Institute of Radio Engineers*, No. 331, November 1945.
8   John Gunther, "They Get the Message Through: The Story of the Signal Corps," *Redbook* (New York) Volume 80, Issue 2, (December 1942): 40–41, 100–102.
9   Melissa Kozlowski (Ziobro), "Winged warriors make a 'valiant' comeback at the theater," *Monmouth Message*, August 22, 2005.
10  Melissa Ziobro, "Gender Integration of the Army Advanced at Fort Monmouth," *Monmouth Message*, March 9, 2007; Melissa Ziobro, "Skirted Soldiers: The Women's Army Corps and Gender Integration of the US Army during WWII," *On Point: The Journal of Army History*, Volume 17, Number 4 (Spring 2012), 36–43.
11  Melissa Ziobro, "Fort Monmouth Drafted Local Sites, Including Famed Asbury Park Convention Hall," *Monmouth Message,* May 25, 2007.
12  Melissa Ziobro, "The Friendly Occupation of Asbury Park," *Garden State Legacy*, June 2020, Issue 48.
13  Marge Bramley interviewed by Douglas Aumack, July 26, 2000. Courtesy Monmouth County Library.
14  "War Wives," *Good Housekeeping* (New) Volume 115, Issue 4, (October 1942): 49–57.
15  "39 Italians in Work Unit Visit City, Run Gantlet of Total Indifference," *New York Times*, February 5, 1945; *A Concise History of Fort Monmouth, NJ, and the US Army Communications-Electronics Life Cycle Management Command* (Fort Monmouth, NJ: Visual Information Services, 2009), 26.
16  Marge Bramley interviewed by Douglas Aumack, July 26, 2000. Courtesy Monmouth County Library.
17  "Plenty of Cigarets (stet) and Butter Keeps Monmouth's PWs Happy, Says Report," *The Asbury Park Evening Press*, January 10, 1945.
18  "To Be Wed in Rome," *Asbury Park Evening Press*, April 24, 1946.
19  "Army Returns Signal School to Monmouth; Camp Crowder Training To End, Enlisted Men's School Reinstated at Shore Post," *Asbury Park Evening Press*, February 22, 1946.
20  William R. Stevenson, *Miniaturization and Microminiaturization of Army Communications-Electronics, 1946–1964* (Fort Monmouth: US Army Electronics

Command, 1966); Max L. Marshall, *The Story of the US Army Signal Corps* (New York: Franklin Watts, Inc., 1965).
21  Project Diana is also discussed in Chapter 6. *The History of Fort Monmouth, 1917–1953*, prepared by the Historical Branch, G3, Fort Monmouth, NJ, 1953. Courtesy US Army Communications-Electronics Command Historical Office, Aberdeen Proving Ground, Maryland.
22  Dickey Meyer, "RADAR, Today's Miracle," *Seventeen* (New York) Volume 6, Issue 1, (Jan 1947): 70–71, 113.

## Chapter 5 "Where the Army Signal Corps Thinks Out Some of the Nation's Crucial Defenses": Cold War Battleground

1  Peter Kihss, "Monmouth Security Foes Antedate McCarthy Visit," *New York Times*, January 11, 1954.
2  National Archives and Records Administration, "US Enters the Korean Conflict," nara.gov, https://www.archives.gov/education/lessons/korean-conflict#background. Accessed June 3, 2023. Department of Veterans Affairs, "America's Wars," va.gov, May 21, 2021, fs_americas_wars.pdf (va.gov). Accessed June 3, 2023.
3  Thomas Daniels interviewed by Robert Johnson, Jr., June 21–24, 1993. Courtesy US Army Communications-Electronics Command Historical Office, Aberdeen Proving Ground, Maryland.
4  William R. Stevenson, *Miniaturization and Micro-miniaturization of Army Communications-Electronics 1946–1964* (Fort Monmouth: US Army Electronics Command, 1966; Max L. Marshall, *The Story of the US Army Signal Corps* (New York: Franklin Watts, Inc., 1965).
5  "Radar Traces Projectiles, Locates Mortar," *Popular Mechanics* Vol. 103, No. 6 (June 1955), 123.
6  *Signal Corps Engineering Laboratories Annual Review*, 1951. Courtesy US Army Communications-Electronics Command Historical Office, Aberdeen Proving Ground, Maryland.
7  William R. Stevenson, *Miniaturization and Microminiaturization of Army Communications-Electronics, 1946–1964* (Fort Monmouth: US Army Electronics Command, 1966).
8  William R. Stevenson, *Miniaturization and Microminiaturization of Army Communications-Electronics, 1946–1964* (Fort Monmouth: US Army Electronics Command, 1966), 242.
9  Anita Impellizeri, "The Deal Test Site," National Register of Historic Places Registration Form, 1981. Courtesy US Army Communications-Electronics Command Historical Office, Aberdeen Proving Ground, Maryland.
10  Melissa Ziobro, "Our Deal Area First to Record Sputnik Signals," *Monmouth Message,* October 19, 2007.
11  Harold Zahl, *Electronics Away or Tales of a Government Scientist* (New York: Vantage Press, 1969), 120–125.

12  "Vanguard 1," NASA.gov, https://nssdc.gsfc.nasa.gov/nmc/spacecraft/display.action?id=1958-002B. Accessed May 6, 2023. See also Max Marshall, *The Story of the US Army Signal Corps* (New York: Franklin Watts, Inc., 1965); *Historical Sketch of the United States Army Signal Corps 1860–1967*, prepared by the Signal School Historical Office, 1967, at Fort Monmouth, NJ and courtesy US Army Communications-Electronics Command Historical Office, Aberdeen Proving Ground, Maryland.

13  "Army Builds $500,000 Chapel at Fort, Other Projects Planned," *Asbury Park Sunday Press*, January 15, 1961; Max Marshall, *The Story of the US Army Signal Corps* (New York: Franklin Watts, Inc., 1965); *Historical Sketch of the United States Army Signal Corps 1860–1967*, prepared by the Signal School Historical Office, 1967, at Fort Monmouth, NJ and courtesy US Army Communications-Electronics Command Historical Office, Aberdeen Proving Ground, Maryland.

14  Sputnik was launched October 4, 1957.

15  This was the aforementioned Explorer project, headed by Wernher von Braun.

16  The Vanguard Program had three failed launch attempts before sending the *Vanguard I* satellite into orbit on March 17, 1958. Operating on solar-powered batteries, it was still transmitting after three years in orbit.

17  John Cittadino interviewed by Wendy Rejan, January 9, 2009. Courtesy US Army Communications-Electronics Command Historical Office, Aberdeen Proving Ground, Maryland.

18  Ibid.

19  Ibid.

20  "Vanguard 1," NASA.gov, https://nssdc.gsfc.nasa.gov/nmc/spacecraft/display.action?id=1958-002B. Accessed May 6, 2023. See also Max Marshall, *The Story of the US Army Signal Corps* (New York: Franklin Watts, Inc., 1965); *Historical Sketch of the United States Army Signal Corps 1860–1967*, prepared by the Signal School Historical Office, 1967, at Fort Monmouth, NJ and courtesy US Army Communications-Electronics Command Historical Office, Aberdeen Proving Ground, Maryland.

21  "Army Builds $500,000 Chapel at Fort, Other Projects Planned," *Asbury Park Sunday Press*, January 15, 1961; Max Marshall, *The Story of the US Army Signal Corps* (New York: Franklin Watts, Inc., 1965); *Historical Sketch of the United States Army Signal Corps 1860–1967*, prepared by the Signal School Historical Office, 1967, at Fort Monmouth, NJ and courtesy US Army Communications-Electronics Command Historical Office, Aberdeen Proving Ground, Maryland.

22  Max Marshall, *The Story of the US Army Signal Corps* (New York: Franklin Watts, Inc., 1965); *Historical Sketch of the United States Army Signal Corps 1860–1967*, prepared by the Signal School Historical Office, 1967, at Fort Monmouth, NJ and courtesy US Army Communications-Electronics Command Historical Office, Aberdeen Proving Ground, Maryland.

23 Melissa Ziobro, "Early Signal Corps satellite programs remembered," *Monmouth Message,* April 19, 2010.
24 John Cittadino interviewed by Wendy Rejan, January 9, 2009. Courtesy US Army Communications-Electronics Command Historical Office, Aberdeen Proving Ground, Maryland.
25 "Army Builds $500,000 Chapel at Fort, Other Projects Planned," *Asbury Park Sunday Press,* January 15, 1961.
26 Celebrating 50th Anniversary of the US Television Infrared Observation Satellite; Congressional Record Vol. 156, No. 65 (House - May 4, 2010).
27 "Six Students Fort's Guests," *Asbury Park Evening Press,* December 2, 1959; *A Century of US Army Signals* (Fort Monmouth, NJ), 16. Courtesy US Army Communications-Electronics Command Historical Office, Aberdeen Proving Ground, Maryland.
28 "Army Builds $500,000 Chapel at Fort, Other Projects Planned," *Asbury Park Sunday Press,* January 15, 1961; Max Marshall, *The Story of the US Army Signal Corps* (New York: Franklin Watts, Inc., 1965).
29 Charles C. Smith interviewed by Frances Prokop, July 20, 2017. Courtesy Library of Congress Veterans History Project, Washington, DC.
30 "Miniature Radar," *Asbury Park Evening Press,* May 19, 1962; William R. Stevenson, *Miniaturization and Microminiaturization of Army Communications-Electronics, 1946–1964* (Fort Monmouth: US Army Electronics Command, 1966).
31 Melissa Kozlowski (Ziobro), "Winged warriors make a 'valiant' comeback at the theater," *Monmouth Message,* August 22, 2005.
32 Also referred to as Operation Paperclip; related to Project Overcoat. For more on Project Paperclip generally, see *Operation Paperclip: The Secret Intelligence Program that Brought Nazi Scientists to America* by Annie Jacobsen. Fred Carl, founder of the InfoAge Science and History Museums at the Camp Evans National Historic Landmark, is currently conducting more research into the experience of the Fort Monmouth/Camp Evans Project Paperclip personnel.
33 Harold Zahl, *Electrons Away or Tales of a Government Scientist* (New York: Vantage Press, 1968), 108–109.
34 E. Burke Maloney, "German Scientists at Fort Monmouth Save US Millions; Further Research," *Asbury Park Sunday Press,* November 30, 1947.
35 Numbers vary slightly from source to source. Richard C. Halverson, "Reflections of an Ex-German Scientists," *Asbury Park Press,* October 28, 1984.
36 Richard C. Halverson, "Reflections of an Ex-German Scientists," *Asbury Park Press,* October 28, 1984.
37 Ibid.
38 Memorandum to Lieutenant Colonel W. M. Young, "Information About German Scientists," June 30, 1948. Courtesy Monmouth County Historical Association.
39 Harold Zahl, *Electrons Away or Tales of a Government Scientist* (New York: Vantage Press, 1968), 108–109.

40  "Questions and Answers During Ninth Day's Hearing in Army-McCarthy Controversy," *Philadelphia Inquirer*, May 5, 1954; "Rebecca Raines, "The Cold War Comes to Fort Monmouth: Senator Joseph R. McCarthy and the Search for Spies in the Signal Corps," *Army History*, No. 44 (Spring 1998), 8–16.
41  Harvey Lasky interviewed by Wendy Rejan, May 12, 2008. Courtesy US Army Communications-Electronics Command Historical Office, Aberdeen Proving Ground, Maryland.
42  "McCarthy: Public Hearings Essential," *Asbury Park Sunday Press*, November 15, 1953.
43  Peter Kihss, "Monmouth Security Foes Antedate McCarthy Visit," *New York Times*, January 11, 1954; David M. Oshinsky, *A Conspiracy So Immense: The World of Joe McCarthy* (New York: Free Press, 1983); Rebecca Raines, "The Cold War Comes to Fort Monmouth: Senator Joseph R. McCarthy and the Search for Spies in the Signal Corps," *Army History*, No. 44 (Spring 1998), 8–16.
44  Dr. William Ryan interviewed by Melissa Ziobro, August 8, 2008. Courtesy US Army Communications-Electronics Command Historical Office, Aberdeen Proving Ground, Maryland.
45  Milton Friedrich Langer interviewed by Walter Diambri, November 11, 2015. Courtesy Library of Congress Veterans History Project, Washington, DC.
46  Leo Geisler interviewed by Ian Girard, date unknown. Courtesy Library of Congress Veterans History Project, Washington, DC.

## Chapter 6 "The Black Brain Center of the United States": Dr. Walter McAfee and His Colleagues Break Barriers at Fort Monmouth

1  This chapter is adapted from the author's article in the Summer 2022 issue of *NJ Studies: An Interdisciplinary Journal*. Used with permission.
2  Walter S. McAfee interviewed by Professor Robert Johnson Jr., February 6, 1994. Courtesy US Army Communications-Electronics Command Historical Office, Aberdeen Proving Ground, Maryland.
3  Wiley College, founded in 1873 in Marshall, Texas, is a Historically Black College (HBCU). See more at https://www.wileyc.edu/the-history-of-wiley-college/.
4  Walter S. McAfee interviewed by Professor Robert Johnson Jr., February 6, 1994. Courtesy US Army Communications-Electronics Command Historical Office, Aberdeen Proving Ground, Maryland.
5  Honoring Dr. Walter S. McAfee; Congressional Record Vol. 165, No. 143 (Extensions of Remarks - September 9, 2019).
6  George Morris interviewed by Melissa Ziobro, March 2, 2021. Courtesy Monmouth Memories Oral History Program.
7  Michelle Drobik, email message to author, June 4, 2021.
8  Walter S. McAfee interviewed by Professor Robert Johnson Jr., February 6, 1994. Courtesy US Army Communications-Electronics Command Historical Office, Aberdeen Proving Ground, Maryland.
9  Ibid.

10  Walter S. McAfee interviewed by Professor Robert Johnson Jr., February 6, 1994. Courtesy US Army Communications-Electronics Command Historical Office, Aberdeen Proving Ground, Maryland.
11  Gloria Stravelli, "Army research facility once known as the 'Black Brain Center,'" https://archive.centraljersey.com/2003/03/27/forts-gate-swung-open-to-black-scientists/. Accessed April 4, 2023.
12  Albert C. Johnson interviewed by Carol Stokes, December 30, 1995. Courtesy US Army Communications-Electronics Command Historical Office, Aberdeen Proving Ground, Maryland; Melissa Ziobro, "COL Albert C. Johnson," *On Point: The Journal of Army History*, Winter 2019 Vol. 24 No. 3.
13  Erik Larsen, "A look back at Wall Township's racist past," *Asbury Park Press*, June 17, 2018, https://www.app.com/story/news/history/erik-larsen/2018/06/17/look-back-wall-townships-racist-past/704099002/. Accessed February 2, 2022.
14  Walter S. McAfee interviewed by Professor Robert Johnson Jr., February 6, 1994. Courtesy US Army Communications-Electronics Command Historical Office, Aberdeen Proving Ground, Maryland.
15  Ibid.
16  Ibid.
17  Ibid.
18  Today, the land is home to the InfoAge Science History Learning Center at the Camp Evans National Historic Landmark. See more at www.infoage.org.
19  Collin Makamsom, "Project Diana: To The Moon And Back," National WWII Museum, https://www.nationalww2museum.org/war/articles/project-diana-moon-and-soviet-union. Accessed February 4, 2022.
20  Walter S. McAfee interviewed by Professor Robert Johnson Jr., February 6, 1994. Courtesy US Army Communications-Electronics Command Historical Office, Aberdeen Proving Ground, Maryland.
21  Jack Gould, "Contact With Moon Achieved By Radar In Test By The Army: An Echo From The Moon Is Recorded," *New York Times*, January 25, 1946; "The First Moon Contact," *Asbury Park Press*, August 29, 1971; Collin Makamsom, "Project Diana: To The Moon And Back," National WWII Museum, https://www.nationalww2museum.org/war/articles/project-diana-moon-and-soviet-union. Accessed February 4, 2022.
22  "Camp Evans Scientists Open Vast New Field with Radar Moon Contact," *Asbury Park Press*, January 25, 1946, 1.
23  "The First Moon Contact," *Asbury Park Press*, August 29, 1971; "Wall, County Officials Praise Diana Workers," *Red Bank Register*, January 12, 1971, 1.
24  Walter S. McAfee interviewed by Professor Robert Johnson Jr., February 6, 1994. Courtesy US Army Communications-Electronics Command Historical Office, Aberdeen Proving Ground, Maryland.
25  Ibid.
26  Ibid.

27 "Announce 50 Rosenwald Fellowships; 31 Go to Negroes for Important Projects," *The Daily Bulletin*, May 21, 1946; "Rosenwald Scholarships Announced For Fifty," *Detroit Tribune*, May 25, 1946.
28 "Area Scientist Wins Fellowship," *Red Bank Register*, November 8, 1956, 46; "Monmouth Physicist Fellowship Winner," *Asbury Park Evening Press*, November 2, 1956, 17.
29 Rebecca Raines, "The Cold War Comes to Fort Monmouth: Senator Joseph R. McCarthy and the Search for Spies in the Signal Corps," *Army History*, No. 44 (Spring 1998), 8–16. The McCarthy period is discussed in greater detail in Chapter 5.
30 Walter S. McAfee interviewed by Professor Robert Johnson Jr., February 6, 1994. Courtesy US Army Communications-Electronics Command Historical Office, Aberdeen Proving Ground, Maryland.
31 *Asbury Park Evening Press*, "3 Signal Scientists Honored," September 12, 1961, 12.
32 "Shore Reacts with Awe," *Asbury Park Press*, July 21, 1969, 30.
33 "McAfee named to GS-16 Post," *Red Bank Register*, January 18, 1971, 13; "Wall, County Officials Praise Diana Workers," *Red Bank Register*, January 12, 1971, 1; "Dr. McAfee Gets High ECOM Post," *Asbury Park Press*, January 19, 1971, 23; "Dr. McAfee's Appointment," *Asbury Park Press*, January 24, 1971, 15.
34 "Dr. McAfee Gets High ECOM Post," *Asbury Park Press*, January 19, 1971, 23; "Dr. McAfee's Appointment," *Asbury Park Press*, January 24, 1971, 15.
35 "Scientist helped boost US into space," *Asbury Park Press*, January 16, 1985, 4.
36 Walter S. McAfee interviewed by Professor Robert Johnson Jr., February 6, 1994. Courtesy US Army Communications-Electronics Command Historical Office, Aberdeen Proving Ground, Maryland.; "The Negro Motorist Green Book," Smithsonian Digital Volunteers: Transcription Center, https://transcription.si.edu/project/7955#:~:text=The%20Negro%20Motorist%20Green%20Book%20was%20a%20guidebook%20for%20African,that%20served%20African%20Americans%20patrons. For further reading on the *Green Book*, see Mia Bay, *Traveling Black: A Story of Race and Resistance* (Belknap Press, 2021) and Gretchen Sorin, *Driving While Black: African American Travel and the Road to Civil Rights* (Liveright, 2020).
37 "Scientist helped boost US into space," *Asbury Park Press*, January 16, 1985, 4; "Senator to Speak at College Graduation," *Asbury Park Press*, April 25, 1985, A14.
38 "College Fund Given By Fort Personnel," *Asbury Park Evening Press*, August 9, 1946, 5.
39 "Math Club Meets At College Wednesday," *Asbury Park Sunday Press*, April 30, 1961, 8; "Door Is Opened to Needy Students," *Asbury Park Sunday Press*, April 26, 1964, 5; "Models," *Asbury Park Press*, February 16, 1986, 4.
40 "Brookdale Trustee Candidate is Eyed," *Red Bank Register*, November 12, 1969, 5.
41 "The New College Trustee," *Red Bank Register*, January 22, 1970, 6.

42  "Brookdale Board Post Goes to Fort Scientist," *Red Bank Register*, January 20, 1970, 13; "Litwin Sworn to Brookdale Board," *Red Bank Register*, November 21, 1975, 29; "Irwin at Helm," *Asbury Park Press*, January 3, 1970, 11; "Litwin Choice to Head Trustees at Brookdale College," *Asbury Park Press*, November 21, 1980, 44; Doris Kulman, "Brookdale faculty to launch class boycott," *The Daily Register*, April 12, 1977, 1; Doris Kulman, "Board remains silent as complaints pile up," *The Daily Register*, April 29, 1977, 1; Walter S. McAfee, "A Letter from the Board of Trustees to the People of Monmouth County," *Red Bank Register*, September 18, 1977, 46; "Brookdale's History," https://www.brookdalecc.edu/about/about-the-college/brookdales-history-2/. Accessed February 4, 2022.
43  "Socratic Society to Talk Science," *Asbury Park Press*, April 23, 1961, 4; "Dr. McAfee Gets High ECOM Post," *Asbury Park Press*, January 19, 1971, 23; "Scientist helped boost US into space," *Asbury Park Press*, January 16, 1985, 4.
44  "Installation Saturday for Sigma Pi Sigma," *Red Bank Register*, January 25, 1967, 31.
45  "George Morris interviewed by Melissa Ziobro," March 2, 2021. Courtesy Monmouth Memories Oral History Program.
46  Ron Johnson interviewed by Melissa Ziobro," April 6, 2021. Courtesy Monmouth Memories Oral History Program.
47  Gary Barnett interviewed by Vincent Sauchelli, December 14, 2020. Courtesy Monmouth Memories Oral History Program.
48  John Tranchina interviewed by Melissa Ziobro, February 22, 2021. Courtesy Monmouth Memories Oral History Program.
49  George Morris interviewed by Melissa Ziobro," March 2, 2021. Courtesy Monmouth Memories Oral History Program.
50  *Asbury Park Evening Press*, "3 Signal Scientists Honored," September 12, 1961, 12; "3 at Fort Monmouth Get Army Awards," *New York Times*, September 17, 1961, 124; *Asbury Park Evening Press*, "Awards Fete Set by NBPW Club," May 12, 1971, 24; "4 to Receive Brotherhood Awards," *Asbury Park Press*, August 19, 1976, 7; "Senator to speak at college graduation," *Asbury Park Press*, April 25, 1985, 14; "Borough confers a place in history," *Asbury Park Press*, August 21, 1997, 18.
51  John Harnes, "CECOM building dedication honors Space Age pioneer," *Asbury Park Press*, July 29, 1997, 6.
52  Tribute To Walter S. McAfee; Congressional Record Vol. 143, No. 144 (Extensions of Remarks - October 23, 1997).
53  Ibid.
54  Kristen Kushiyama, "Legacy of NJ scientist transitioned to MD," October 6, 2011, https://www.army.mil/article/65848/legacy_of_nj_scientist_transitioned_to_md. Accessed February 2, 2022.
55  Greg Mahall, "First African-American enters AMC Hall of Fame," *The Redstone Rocket*, October 12, 2015.
56  "AMC Hall of Fame," https://www.amc.army.mil/Organization/History/Hall-of-Fame/. Accessed December 5, 2021.

57  Greg Mahall, "First African-American enters AMC Hall of Fame," https://www.army.mil/article/157432/First_African_American_enters_AMC_Hall_of_Fame/. Accessed January 12, 2022.
58  "Post Office Building Renaming Ceremony: Belmar Post Office building to be renamed in honor of former South Belmar resident Dr. Walter S. McAfee," https://about.usps.com/newsroom/local-releases/nj/2019/0828ma-po-building-renaming-ceremony.htm. Accessed February 4, 2022.
59  Cathy Goetz, "Belmar Post Office Dedication: Walter S. McAfee Honored for Blazing a Path to Space Age," https://www.tapinto.net/towns/belmar-slash-lake-como/sections/honors-and-achievements/articles/belmar-post-office-dedication-walter-s-mcafee-honored-for-blazing-a-path-to-space-age. Accessed February 1, 2022.
60  Susanne Cervanka, "Belmar post office renamed after Walter McAfee, who helped get us to the moon," *Asbury Park Press*, September 2, 2019.
61  "Honoring the Legacy of Walter S. McAfee '85HN," https://www.monmouth.edu/university-advancement/mcafee/. Accessed October 12, 2021.
62  Matthew Cutillo, "Dr. Walter McAfee's Legacy Honored by University Faculty," *The Outlook*, https://outlook.monmouth.edu/2021/02/dr-walter-mcafee-s-legacy-honored-by-university-faculty/. Accessed October 21, 2021.

## Chapter 7 "The Eyes, Ears, and Voice of the Fighting Man": The Vietnam War and Its Aftermath

1  The title of the chapter is derived from the title of a 1960s military film about the mission of the US Army Electronics Command at Fort Monmouth, available via the US Army Communications-Electronics Command Historical Office, Aberdeen Proving Ground, Maryland.
2  Historian, US Army Electronics Command, *US Army Electronics Command Logistics Support of the Army in Southeast Asia 1965—1970* (Fort Monmouth: US Army Electronics Command, 1973), 1. Lotz would later be promoted for Major General and serve as Commanding General of ECOM and Fort Monmouth from September 1969—May 1971.
3  Max Cleland interviewed by David Taylor and Edwin Perry, undated. Courtesy Library of Congress Veterans History Project.
4  For more on these reorganizations, see James E. Hewes, Jr., *From Root To McNamara: Army Organization and Administration* (Washington, DC: US Army Center of Military History, 1975), Chapter 10.
5  "Army to Base Electronics Unit at Fort," *Asbury Park Evening Press*, May 24, 1962.
6  "New Command Job Impact Uncertain," *Asbury Park Evening Press*, May 27, 1962.
7  Rebecca Robbins Raines, *Getting the Message Through: A Branch History of the US Army Signal Corps* (Washington, DC: US Army Center of Military History, 1996), Chapters 9 and 10; Helen Phillips, *United States Army Signal School, 1919–1967*.

Courtesy US Army Communications-Electronics Command Historical Office, Aberdeen Proving Ground, Maryland.
8 "Signal Corps School," *Asbury Park Evening Press*, March 6, 1971.
9 William Ryan interviewed by Melissa Ziobro, August 8, 2008. US Army Communications-Electronics Command Historical Office, Aberdeen Proving Ground, Maryland.
10 "Signaling the End of An Era: School to Hold Last Gradation," *Monmouth Message*, June 16, 1976.
11 James V. Gannon interviewed by William Strong, February 25, 1987. Courtesy US Army Communications-Electronics Command Historical Office, Aberdeen Proving Ground, Maryland.
12 The most comprehensive analysis of Fort Monmouth's support to operations in Southeast Asia is *US Army Electronics Command Logistics Support of the Army in Southeast Asia 1965–1970* (Fort Monmouth: US Army Electronics Command, 1973).
13 As often as possible, this books seeks to avoid military jargon, but AN/PRC stands for Army Navy/Portable Radio Communications and AN/VRC stands for Army Navy Vehicle Radio Communications.
14 AN/PRR stands for Army Navy/Portable Radio Receiver.
15 AN/PRT stands for Army Navy/Portable Radio Transmitter.
16 Historian, US Army Electronics Command, *US Army Electronics Command Logistics Support of the Army in Southeast Asia 1965–1970* (Fort Monmouth: US Army Electronics Command, 1973), 83.
17 AN/MPQ stands for Army/Navy Mobile Position Finder Special Purpose.
18 Historian, US Army Electronics Command, *US Army Electronics Command Logistics Support of the Army in Southeast Asia 1965–1970* (Fort Monmouth: US Army Electronics Command, 1973), 50.
19 "Mortar Locator Used in Vietnam ECOM Developed," *Monmouth Message*, December 1, 1966.
20 Historian, US Army Electronics Command, *US Army Electronics Command Logistics Support of the Army in Southeast Asia 1965–1970* (Fort Monmouth: US Army Electronics Command, 1973), 70–82.
21 "New, Faster Meteorological Data Sounding System Under Development by Electronics Command," *Monmouth Message*, November 10, 1965.
22 "New Satellite Ground Terminal Operation Nears," *Monmouth Message*, December 1, 1965.
23 "Work on Night Vision is Assigned to ECOM," *Monmouth Message*, November 10, 1965.
24 "West Deal Man Wins Top Award," *Asbury Park Evening Press,* December 20, 1968.
25 Historian, US Army Electronics Command, *US Army Electronics Command Logistics Support of the Army in Southeast Asia 1965–1970* (Fort Monmouth: US Army Electronics Command, 1973), 3.
26 Rockwell C. Webb interviewed by Paul Plummer and Tina Rodriguez, March 22, 2017. Library of Congress Veterans History Project.

27 "ECOM Seeking Civilians for Vacancies in Vietnam," *Monmouth Message*, November 10, 1965.
28 Rick Amsterdam email to Melissa Ziobro, August 3, 2023. Author's collection.
29 Michael Coale email to Melissa Ziobro, August 3, 2023. Author's collection.
30 Ron Wentworth email to Melissa Ziobro, August 3, 2023. Author's collection.
31 This letter is archived in the extensive pigeon collection of the US Army Communications-Electronics Command Historical Office, Aberdeen Proving Ground, Maryland.
32 Melissa Ziobro, "CECOM Predecessor's Birthday Recalled," *Monmouth Message*, July 30, 2010; "Two Fort Monmouth Commands to be Merged," *Asbury Park Press*, December 19, 1980.
33 Ellen Carroll, "Forts Electronic Wizardry was There When It Counted," *Asbury Park Press*, August 1, 1982.
34 Edward L. Walsh, "Fort Monmouth: Surviving Four Wars," *Asbury Park Press*, May 4, 1980.
35 Ellen Carroll, "Fort Monmouth Lived to See a New Horizon," *Asbury Park Press*, August 1, 1982.

# Chapter 8 "We Were Able to Control our Force Much Better than the Enemy Controlled His": Operations Desert Shield and Desert Storm

1 James V. Gannon interviewed by William Strong, February 25, 1987. Courtesy US Army Communications-Electronics Command Historical Office, Aberdeen Proving Ground, Maryland.
2 George Akin interviewed by William Strong and Julius Simchick, April 4, 1988. Courtesy US Army Communications-Electronics Command Historical Office, Aberdeen Proving Ground, Maryland.
3 Raymond E. B. Ketchum II interviewed by William Strong and Julius Simchick, July 10, 1987. Courtesy US Army Communications-Electronics Command Historical Office, Aberdeen Proving Ground, Maryland.
4 James V. Gannon interviewed by William Strong, February 25, 1987. Courtesy US Army Communications-Electronics Command Historical Office, Aberdeen Proving Ground, Maryland.
5 Raymond E. B. Ketchum II interviewed by William Strong and Julius Simchick, July 10, 1987. Courtesy US Army Communications-Electronics Command Historical Office, Aberdeen Proving Ground, Maryland.
6 Ibid.
7 Ibid.
8 Melissa Ziobro, "Cold War Competition Heats up Innovation at Fort Monmouth: Part III," *Monmouth Message*, November 9, 2007.
9 Ibid.
10 Ibid.

11 Ibid.
12 Ibid.
13 "Army Picks ITT Equipment," *Asbury Park Press*, July 30, 1983; *Asbury Park Press*, October 20, 1987.
14 National Museum of the US Army, "Cellular Communications," https://www.thenmusa.org/armyinnovations/cellularcommunications/. Accessed June 5, 2023.
15 Melissa Ziobro, "Cold War Competition Heats up Innovation at Fort Monmouth: Part III," *Monmouth Message*, November 9, 2007.
16 US Air Force, "E-8C Joint Stars," https://www.af.mil/About-Us/Fact-Sheets/Display/Article/104507/e-8c-joint-stars/. Accessed June 5, 2023.
17 Davind C. Morrison, "In General, It's Just Pentagonese," *Asbury Park Press*, July 17, 1986.
18 Bobby Mayfield interviewed by Ruth Evans, November 25, 2005. Courtesy Library of Congress Veterans History Project.
19 United States Congress House Committee on Appropriations Subcommittee on Department of Defense, *Department of Defense Appropriations for Fiscal Year 1992: Research, Development, Test and Evaluation: Army* (Washington, DC: Government Printing Office, 1991), 36, 439.
20 Dr. Richard Bingham, *CECOM and the War for Kuwait August 1990–March 1991* (Fort Monmouth: US Army Communications-Electronics Command, 1994).
21 Melissa Ziobro, "CECOM Building Demolished," *Monmouth Message*, July 17, 2009.

## Chapter 9 "It Has Definitely Left Some Lasting Marks that Aren't Necessarily Easy to Get Over": Support of the Global War on Terror

1 National Commission on the Terrorist Attacks Upon the United States, *The 9/11 Commission Report: Final Report of the National Commission on Terrorist Attacks Upon the United States* (Washington, DC: Government Printing Office, 2004), Executive Summary.
2 Mike Ruane interviewed by Kelly Dender, December 3, 2021. Courtesy 9/11 Oral History Project, Monmouth University.
3 Fort Monmouth employee interviewed by Austin Wagner, November 11, 2021. Courtesy 9/11 Oral History Project, Monmouth University.
4 Ibid.
5 Ibid.
6 Mike Ruane interviewed by Kelly Dender, December 3, 2021. Courtesy 9/11 Oral History Project, Monmouth University.
7 Tom Braumuller interviewed by Melissa Ziobro, December 5, 2023. Courtesy Bruce Springsteen Archives and Center for American Music, Monmouth University.
8 Fort Monmouth employee interviewed by Melissa Ziobro, July 29, 2021. Courtesy 9/11 Oral History Project, Monmouth University.
9 Ibid.

10   The Terrorist Attack and Tragedy at the World Trade Center; Congressional Record Vol. 147, No. 131 (House - October 3, 2001).
11   C4ISR stood for Command, Control, Communications, Computers, Intelligence, Surveillance, and Reconnaissance. As we have discussed the names used for the missions located at Fort Monmouth and its satellites changed frequently over the years due to Army reorganizations. The term "Team C4ISR" was en vogue at this time, referring primarily to the Communications-Electronics Command (CECOM) and the Program Executive Offices (PEO) discussed in Chapter 8.
12   Fort Monmouth employee interviewed by Wendy Rejan, January 17, 2008. Courtesy US Army Communications-Electronics Command Historical Office, Aberdeen Proving Ground, Maryland.
13   Fort Monmouth employee interviewed by Wendy Rejan, June 17, 2003. Courtesy US Army Communications-Electronics Command Historical Office, Aberdeen Proving Ground, Maryland.
14   Historical Office Staff, *A Concise History of Fort Monmouth, New Jersey and the US Army CECOM Life Cycle Management Command* (Fort Monmouth, NJ, 2007), 74.
15   Fort Monmouth employee interviewed by Wendy Rejan, June 17, 2003. Courtesy US Army Communications-Electronics Command Historical Office, Aberdeen Proving Ground, Maryland.
16   Fort Monmouth employee interviewed by Wendy Rejan, July 22, 2003. Courtesy US Army Communications-Electronics Command Historical Office, Aberdeen Proving Ground, Maryland.
17   Historical Office Staff, *A Concise History of Fort Monmouth, New Jersey and the US Army CECOM Life Cycle Management Command* (Fort Monmouth, NJ, 2007), 69.
18   Historical Office Staff, *A Concise History of Fort Monmouth, New Jersey and the US Army CECOM Life Cycle Management Command* (Fort Monmouth, NJ, 2007), 81.
19   Historical Office Staff, *A Concise History of Fort Monmouth, New Jersey and the US Army CECOM Life Cycle Management Command* (Fort Monmouth, NJ, 2007), 86.
20   Fort Monmouth employee interviewed by Wendy Rejan, July 22, 2003. Courtesy US Army Communications-Electronics Command Historical Office, Aberdeen Proving Ground, Maryland.
21   Fort Monmouth employee interviewed by Wendy Rejan, November 7, 2007. Courtesy US Army Communications-Electronics Command Historical Office, Aberdeen Proving Ground, Maryland.
22   Fort Monmouth employee interviewed by Wendy Rejan, June 18, 2007. Courtesy US Army Communications-Electronics Command Historical Office, Aberdeen Proving Ground, Maryland.

## Chapter 10 The Army Goes Rolling Along: Behind the Scenes of a Base Closure

1   "The Army Goes Rolling Along" is the official song of the United States Army.
2   "Save the Fort," *Asbury Park Press*, July 8, 1975.

3   "DoD Base Realignment and Closure BRAC Rounds (BRAC 1988, 1991, 1993, 1995 & 2005) Executive Summary Fiscal Year (FY) 2023 Budget Estimates," Program Year 2023, https://comptroller.defense.gov/Portals/45/Documents/defbudget/fy2023/budget_justification/pdfs/05_BRAC/FY2023_BRAC_Overview.pdf. Accessed August 1, 2023.
4   Roberta Wells, "Panel Wants Humans Considered in Environmental Impact Studies," *Asbury Park Press*, July 31, 1990.
5   At the time this book was being prepared, the author also had a book chapter focused solely on the closure of Camp Evans pending for publication in an edited volume.
6   "DoD Base Realignment and Closure BRAC Rounds (BRAC 1988, 1991, 1993, 1995 & 2005) Executive Summary Fiscal Year (FY) 2023 Budget Estimates," Program Year 2023, https://comptroller.defense.gov/Portals/45/Documents/defbudget/fy2023/budget_justification/pdfs/05_BRAC/FY2023_BRAC_Overview.pdf. Accessed August 1, 2023.
7   The full report, outlining the specific recommendations, can be found online. See "2005 Defense Base Closure and Realignment Commission Report to the President," https://www.acq.osd.mil/brac/docs/BRAC-2005-Commission-Report.pdf. Accessed March 6, 2023.
8   National Defense Authorization Act for Fiscal Year 2006; Congressional Record Vol. 151, No. 71 (House - May 25, 2005).
9   As noted in Chapter 9, C4ISR stood for Command, Control, Communications, Computers, Intelligence, Surveillance, and Reconnaissance. As we have discussed the names used for the missions located at Fort Monmouth and its satellites changed frequently over the years due to Army reorganizations. This is the term that was en vogue in 2005, referring primarily to the Communications-Electronics Command (CECOM) and the Program Executive Offices (PEO) discussed in Chapter 8. We should also note that the CECOM was organized into something called the CECOM Life Cycle Management Command for a period during this BRAC era; though that convention has since been dropped.
10  "Behind the News," *Asbury Park Press*, September 7, 2006; "2005 Defense Base Closure and Realignment Commission Report to the President," https://www.acq.osd.mil/brac/docs/BRAC-2005-Commission-Report.pdf. Accessed March 6, 2023.
11  "2005 Defense Base Closure and Realignment omission Report to the President," https://www.acq.osd.mil/brac/docs/BRAC-2005-Commission-Report.pdf. Accessed March 6, 2023.
12  Disapproving the Recommendations of the Defense Base Closure and Realignment Commission; Congressional Record Vol. 151, No. 139 (House - October 27, 2005).
13  Fort Monmouth employee interviewed by Wendy Rejan, January 9, 2009. Courtesy US Army Communications-Electronics Command Historical Office, Aberdeen Proving Ground, Maryland.

14 Fort Monmouth employee interviewed by Wendy Rejan, June 26, 2008. Courtesy US Army Communications-Electronics Command Historical Office, Aberdeen Proving Ground, Maryland.
15 Fort Monmouth employee interviewed by Melissa Ziobro, September 30, 2009. Courtesy US Army Communications-Electronics Command Historical Office, Aberdeen Proving Ground, Maryland.
16 Fort Monmouth employee interviewed by Wendy Rejan, May 30, 2008. Courtesy US Army Communications-Electronics Command Historical Office, Aberdeen Proving Ground, Maryland.
17 Ibid.
18 Fort Monmouth employee Personal Experience Paper, May 13, 2008. Courtesy US Army Communications-Electronics Command Historical Office, Aberdeen Proving Ground, Maryland.
19 Keith Brown, "Fort's Fate Up in the Air," *Asbury Park Press*, June 4, 2005.
20 Military Construction and Veterans Affairs Appropriations Act, 2008; Congressional Record Vol. 153, No. 131 (Senate - September 6, 2007).
21 Keith Brown, "Plan B was Rear-guard Action in Battle for Fort," *Asbury Park Press*, August 26, 2005.
22 "The Closing of Monmouth Base in New Jersey Means Job Losses to The Surrounding Communities," *Voice of America*, October 28, 2009. Courtesy US Army Communications-Electronics Command Historical Office, Aberdeen Proving Ground, Maryland.
23 Letter to the Commission from Maria Gatta, Mayor of Borough of Oceanport, letter, June 28, 2005; (https://digital.library.unt.edu/ark:/67531/metadc16513/m1/2/: accessed October 9, 2023), University of North Texas Libraries, UNT Digital Library, https://digital.library.unt.edu; crediting UNT Libraries Government Documents Department.
24 Keith Brown, "Fort's Last-Gasp Hopes Dashed," *Asbury Park Press*, October 28, 2005.
25 Fort Monmouth employee interviewed by Wendy Rejan, June 18, 2007. Courtesy US Army Communications-Electronics Command Historical Office, Aberdeen Proving Ground, Maryland.
26 Military Construction and Veterans Affairs Appropriations Act, 2008; Congressional Record Vol. 153, No. 131 (Senate - September 6, 2007).
27 Fort Monmouth employee interviewed by Melissa Ziobro, September 23, 2009. Courtesy US Army Communications-Electronics Command Historical Office, Aberdeen Proving Ground, Maryland.
28 Fort Monmouth employee interviewed by Wendy Rejan, June 26, 2008. Courtesy US Army Communications-Electronics Command Historical Office, Aberdeen Proving Ground, Maryland.
29 Fort Monmouth employee interviewed by Wendy Rejan, July 31, 2008. Courtesy US Army Communications-Electronics Command Historical Office, Aberdeen Proving Ground, Maryland.

30  Fort Monmouth employee interviewed by Melissa Ziobro, September 11, 2009. Courtesy US Army Communications-Electronics Command Historical Office, Aberdeen Proving Ground, Maryland.
31  Fort Monmouth employee interviewed by Wendy Rejan, June 26, 2008. Courtesy US Army Communications-Electronics Command Historical Office, Aberdeen Proving Ground, Maryland.
32  Fort Monmouth employee interviewed by Melissa Ziobro, September 30, 2009. Courtesy US Army Communications-Electronics Command Historical Office, Aberdeen Proving Ground, Maryland.
33  Fort Monmouth employee interviewed by Wendy Rejan, May 9, 2008. Courtesy US Army Communications-Electronics Command Historical Office, Aberdeen Proving Ground, Maryland.
34  Military Construction and Veterans Affairs Appropriations Act, 2008; Congressional Record Vol. 153, No. 131 (Senate - September 6, 2007).
35  Fort Monmouth employee interviewed by Wendy Rejan, June 26, 2008. Courtesy US Army Communications-Electronics Command Historical Office, Aberdeen Proving Ground, Maryland.
36  Fort Monmouth employee interviewed by Wendy Rejan, June 18, 2007. Courtesy US Army Communications-Electronics Command Historical Office, Aberdeen Proving Ground, Maryland.
37  Fort Monmouth employee interviewed by Wendy Rejan, November 7, 2007. Courtesy US Army Communications-Electronics Command Historical Office, Aberdeen Proving Ground, Maryland.
38  Fort Monmouth employee interviewed by Melissa Ziobro, September 30, 2009. Courtesy US Army Communications-Electronics Command Historical Office, Aberdeen Proving Ground, Maryland.
39  Fort Monmouth employee interviewed by Melissa Ziobro, June 11, 2009. Courtesy US Army Communications-Electronics Command Historical Office, Aberdeen Proving Ground, Maryland.
40  Fort Monmouth employee interviewed by Wendy Rejan, May 30, 2008. Courtesy US Army Communications-Electronics Command Historical Office, Aberdeen Proving Ground, Maryland.
41  Fort Monmouth employee interviewed by Floyd Hertweck, October 8, 2009. Courtesy US Army Communications-Electronics Command Historical Office, Aberdeen Proving Ground, Maryland.
42  Fort Monmouth employee interviewed by Floyd Hertweck, October 13, 2009. Courtesy US Army Communications-Electronics Command Historical Office, Aberdeen Proving Ground, Maryland.
43  Fort Monmouth employee interviewed by Floyd Hertweck, September 30, 2009.
44  Maryland Department of Business and Economic Development, "2005 BRAC State of Maryland Impact Analysis: 2006–2020 (Executive Summary)," https://commerce.

maryland.gov/Documents/ResearchDocument/ImpactAnalysisExecutiveSummary.pdf. Accessed October 5, 2023.
45 Andricka Thomas and Bob DiMichele, "CECOM Commander Retires after 34 Years of Army Service," https://www.army.mil/article/73346/cecom_commander_retires_after_34_years_of_army_service. Accessed April 25, 2023.
46 Ibid.
47 At the time this book was being prepared, the author also had a book chapter focused solely on the disposition of the Fort Monmouth memorials pending for publication in an edited volume.
48 Aaron Mehta, "30 Years: Base Realignment and Closure—A Bitter Pill," *Defense News*, October 25, 2016.

## Conclusion

1 Dennis Via interviewed by Melissa Ziobro, June 11, 2009. Courtesy US Army Communications-Electronics Command Historical Office, Aberdeen Proving Ground, Maryland.

# Select Bibliography

## Books

Bay, Mia. *Traveling Black: A Story of Race and Resistance* (Belknap Press, 2021).
Clark, Christopher. *The Sleepwalkers: How Europe Went to War in 1914* (New York: Harper Perennial, 2014).
Cunningham, John T. *This is New Jersey* (New Brunswick: Rutgers University Press, 1978).
Cunningham, John T. *New Jersey: A Mirror on America* (Andover: Afton Publishing Company, 1997).
Doughtery, Linda. *The Golden Age of New Jersey Horse Racing* (CreateSpace Independent Publishing Platform, 2016).
Galton, Lawrence, and Harold J. Wheelock. *A History of Fort Monmouth, New Jersey, 1917–1946* (Fort Monmouth: Signal Corps Publication Agency, 1946).
Geffken, Rick, and Muriel J. Smith. *Hidden History of Monmouth County* (The History Press, 2019).
Green, Howard L. *Words That Make New Jersey History* (New Brunswick: Rutgers University Press, 1995).
Hazard, Sharon. *Long Branch in the Golden Age: Tales of Fascinating and Famous People* (The History Press, 2007).
Hewes, James E., Jr. *From Root To McNamara: Army Organization and Administration* (Washington, DC: US Army Center of Military History, 1975).
Liddell Hart, B. H. *History of the First World War* (London: Pan Books, 1972).
Lurie, Maxine, and Richard Veit. *New Jersey: A History of the Garden State* (Rutgers University Press, 2012).
Lurie, Maxine. *Taking Sides in Revolutionary New Jersey: Caught in the Crossfire* (Rutgers University Press, 2022).
Marshall, Max L. *The Story of the US Army Signal Corps* (New York: Franklin Watts, Inc., 1965).
Moss, George H., Jr., *Twice Told Tales: Reflections of Monmouth County's Past* (Sea Bright: Ploughshare Press, 2002).
Moss, George H., Jr., and Karen L. Schnitzspahn. *Victorian Summers at the Grand Hotels of Long Branch, New Jersey* (Sea Bright: Ploughshare Press, 2000).

Pike, Helen C., and Glenn D. Vogel. *Eatontown and Fort Monmouth* (Great Britain: Arcadia Publishing, 1995).
Rejan, Wendy; Chrissie Reilly, and Melissa Ziobro. *A History of Army Communications and Electronics at Fort Monmouth, New Jersey, 1917–2007* (Washington, DC: Government Printing Office, 2008).
Rejan, Wendy. *Images of America: Fort Monmouth* (Arcadia Publishing, 2009).
Robbins Raines, Rebecca. *Getting the Message Through: A Branch History of the US Army Signal Corps* (Washington, DC: US Army Center of Military History, 1996).
Russell, Robert, and Richard Youmans. *Down the Jersey Shore* (New Brunswick: Rutgers University Press, 1993).
Soderlund, Jean. *Lenape Country Delaware Valley Society Before William Penn* (University of Pennsylvania Press, 2014).
Soderlund, Jean. *Separate Paths: Lenapes and Colonists in West New Jersey* (Rutgers University Press, 2022).
Sorin, Gretchen. *Driving While Black: African American Travel and the Road to Civil Rights* (Liveright Publishing Corporation, 2020).
Thompson, George Raynor, and Dixie R. Harris. *The Signal Corps: The Outcome* (Washington, DC: Office of the Chief of Military History, 1966).
Weigley, Russell F. *History of the United States Army* (New York: Macmillan Publishing Company, Incorporated, 1967).
Weigley, Russell F. *Eisenhower's Lieutenants: The Campaigns of France and Germany, 1944–1945* (Bloomington, IN: Indiana University Press, 1981).
Zahl, Harold. *Electronics Away or Tales of a Government Scientist* (New York: Vantage Press, 1969).
Zahl, Harold. *Radar Spelled Backwards* (New York: Vantage Press, 1972).

## Select Newspapers

*Asbury Park Press*
*Asbury Park Evening Press*
*Daily Bulletin*
*Daily Register*
*Detroit Tribune*
*Dots and Dashes*
*Herald Tribune*
*Monmouth Message*
*New York Herald*
*New York Times*
*New York Tribune*
*Philadelphia Inquirer*
*Red Bank Register*

*Signal Corps Message*
*Washington Post*

## Oral History Collections

9/11 Oral History Collection, Monmouth University, West Long Branch, New Jersey.
Bruce Springsteen Archives and Center for American Music at Monmouth University, West Long Branch, New Jersey.
Library of Congress Veterans History Project, Washington, DC.
Monmouth County Library Oral History Collection, Manalapan, New Jersey.
Monmouth Memories Oral History Collection, Monmouth University, West Long Branch, New Jersey.
US Army Communications-Electronics Command Historical Office, Aberdeen Proving Ground, Maryland.

## Archival Collections

The George Moss Collection at the Monmouth University Murry and Leonie Guggenheim Memorial Library, West Long Branch, New Jersey.
The InfoAge Science and History Museums at the Camp Evans National Historic Landmark, Wall Township, New Jersey.
The Monmouth County Historical Association, Freehold, New Jersey.
The US Army Communications-Electronics Command Historical Office, Aberdeen Proving Ground, Maryland.